# Handbook of Superconductivity

# Handbook of Superconductivity

**Edited by**
Victor Ogden

Willford Press
www.willfordpress.com

Published by Willford Press,
118-35 Queens Blvd., Suite 400,
Forest Hills, NY 11375, USA



ISBN: 978-1-68285-497-6

**Cataloging-in-Publication Data**

Handbook of superconductivity / edited by Victor Ogden.
p. cm.
Includes bibliographical references and index.
ISBN 978-1-68285-497-6
1. Superconductivity. 2. Superfluidity. 3. Critical currents. I. Ogden, Victor.
QC611.92 .H36 2018
537.623--dc23

For information on all Willford Press publications
visit our website at www.willfordpress.com

Willford Press

# Contents

# Preface

Superconductors are the materials that are used in transferring and conducting electricity without resistance. Some of the applications of superconductivity are magnetic resonance imaging, magnetic confinement fusion reactors, RF and microwave filters and nuclear magnetic resonance machines. In this book, constant effort has been made to make the understanding of the difficult concepts of Superconductivity, as easy and informative as possible, for the readers. Those in search of information to further their knowledge will be greatly assisted by this book.

To facilitate a deeper understanding of the contents of this book a short introduction of every chapter is written below:

Chapter 1- The phenomenon of zero electrical resistance and magnetic flux field expulsion that occur in materials is known as superconductivity. In most conductors, impurities can cause resistance. When the element is cooled below its critical temperature, a superconductor can be made. The chapter on superconductivity offers an insightful focus, keeping in mind the complex subject matter.

Chapter 2- Diamagnetism, Meissner effect, critical current and field and isotope effect are some of the features of superconductivity. The expulsion of a magnetic field from a superconductor is known as the Meissner effect. The chapter strategically encompasses and incorporates the major components and key concepts of superconductivity, providing a complete understanding.

Chapter 3- The Ginzburg-Landau theory considers the superconducting transition as one of the second order phase transitions. The amplitude $\psi$ is zero in the normal phase above a superconducting transition temperature Tc. It is nonzero below a phase transition into a superconducting state. Superconductivity is best understood in confluence with the major topics listed in the following chapter.

Chapter 4- Cooper pairs are pairs of electrons that bind together at lower temperature. It is fundamental to the property of superconductivity. The topics discussed in the chapter are of great importance to broaden the existing knowledge on superconductivity.

Chapter 5- Tunnel junctions are barriers between two electrically conducting materials. Time-dependent perturbation theory can accurately calculate between different tunneling junctions. The aspects elucidated in this chapter are of vital importance, and provide a better understanding of superconductivity.

Chapter 6- A good value of resistivity of a metal at room temperature is 10-6 ohm cm. At high temperatures, heat capacity tends to a constant value while at low temperatures the heat capacity is a combination of a linear term and a cubic term. The topics discussed in the chapter are of great importance to broaden the existing knowledge on superconductivity.

Finally, I would like to thank the entire team involved in the inception of this book for their valuable time and contribution. This book would not have been possible without their efforts. I would also like to thank my friends and family for their constant support.

**Editor**

# 1

# Understanding Superconductivity

The phenomenon of zero electrical resistance and magnetic flux field expulsion that occur in materials is known as superconductivity. In most conductors, impurities can cause resistance. When the element is cooled below its critical temperature, a superconductor can be made. The chapter on superconductivity offers an insightful focus, keeping in mind the complex subject matter.

## Superconductivity

Superconductivity is a phenomenon of exactly zero electrical resistance and expulsion of magnetic flux fields occurring in certain materials, called superconductors, when cooled below a characteristic critical temperature. It was discovered by Dutch physicist Heike Kamerlingh Onnes on April 8, 1911, in Leiden. Like ferromagnetism and atomic spectral lines, superconductivity is a quantum mechanical phenomenon. It is characterized by the Meissner effect, the complete ejection of magnetic field lines from the interior of the superconductor as it transitions into the superconducting state. The occurrence of the Meissner effect indicates that superconductivity cannot be understood simply as the idealization of *perfect conductivity* in classical physics.

A high-temperature superconductor levitating above a magnet.

The electrical resistance of a metallic conductor decreases gradually as temperature is lowered. In ordinary conductors, such as copper or silver, this decrease is limited by impurities and other defects. Even near absolute zero, a real sample of a normal conductor shows some resistance. In a superconductor; the resistance drops abruptly to zero when the material is cooled below its critical temperature. An electric current flowing through a loop of superconducting wire can persist indefinitely with no power source.

In 1986, it was discovered that some cuprate-perovskite ceramic materials have a critical temperature above 90 K (−183 °C). Such a high transition temperature is theoretically impossible for a conventional superconductor, leading the materials to be termed high-temperature superconductors. The cheaply-available coolant liquid nitrogen boils at 77 K, and thus superconduction at higher temperatures than this facilitates many experiments and applications that are less practical at lower temperatures.

## Classification

There are many criteria by which superconductors are classified. The most common are:

### Response to a Magnetic Field

A superconductor can be *Type I*, meaning it has a single critical field, above which all superconductivity is lost; or *Type II*, meaning it has two critical fields, between which it allows partial penetration of the magnetic field.

### By Theory of Operation

It is *conventional* if it can be explained by the BCS theory or its derivatives, or *unconventional*, otherwise.

### By Critical Temperature

A superconductor is generally considered *high-temperature* if it reaches a superconducting state when cooled using liquid nitrogen – that is, at only $T_c > 77$ K) – or *low-temperature* if more aggressive cooling techniques are required to reach its critical temperature.

### By Material

Superconductor material classes include chemical elements (e.g. mercury or lead), alloys (such as niobium-titanium, germanium-niobium, and niobium nitride), ceramics (YBCO and magnesium diboride), superconducting pnictides (like fluorine-doped LaOFeAs) or organic superconductors (fullerenes and carbon nanotubes; though perhaps these examples should be included among the chemical elements, as they are composed entirely of carbon).

## Elementary Properties of Superconductors

Most of the physical properties of superconductors vary from material to material, such as the heat capacity and the critical temperature, critical field, and critical current density at which superconductivity is destroyed.

On the other hand, there is a class of properties that are independent of the underlying material. For instance, all superconductors have *exactly* zero resistivity to low applied currents when there is no magnetic field present or if the applied field does not exceed a critical value. The existence of these "universal" properties implies that superconductivity is a thermodynamic phase, and thus possesses certain distinguishing properties which are largely independent of microscopic details.

### Zero Electrical DC Resistance

The simplest method to measure the electrical resistance of a sample of some material is to place it in an electrical circuit in series with a current source $I$ and measure the resulting voltage $V$ across the sample. The resistance of the sample is given by Ohm's law as $R = V / I$. If the voltage is zero, this means that the resistance is zero.

Electric cables for accelerators at CERN. Both the massive and slim cables are rated for 12,500 A. *Top:* conventional cables for LEP; *bottom:* superconductor-based cables for the LHC.

Cross section of a preform superconductor rod from abandoned Texas Superconducting Super Collider (SSC).

Superconductors are also able to maintain a current with no applied voltage whatsoever, a property exploited in superconducting electromagnets such as those found in MRI machines. Experiments have demonstrated that currents in superconducting coils can persist for years without any measurable degradation. Experimental evidence points to a current lifetime of at least 100,000 years. Theoretical estimates for the lifetime of a persistent current can exceed the estimated lifetime of the universe, depending on the wire geometry and the temperature.

In a normal conductor, an electric current may be visualized as a fluid of electrons moving across a heavy ionic lattice. The electrons are constantly colliding with the ions in the lattice, and during each collision some of the energy carried by the current is absorbed by the lattice and converted into heat, which is essentially the vibrational kinetic energy of the lattice ions. As a result, the energy carried by the current is constantly being dissipated. This is the phenomenon of electrical resistance and Joule heating.

The situation is different in a superconductor. In a conventional superconductor, the electronic fluid cannot be resolved into individual electrons. Instead, it consists of bound *pairs* of electrons known as Cooper pairs. This pairing is caused by an attractive force between electrons from the exchange of phonons. Due to quantum mechanics, the energy spectrum of this Cooper pair fluid possesses an *energy gap*, meaning there is a minimum amount of energy $\Delta E$ that must be supplied in order to excite the fluid. Therefore, if $\Delta E$ is larger than the thermal energy of the lattice, given by *kT*, where *k* is Boltzmann's constant and *T* is the temperature, the fluid will not be scattered by the lattice. The Cooper pair fluid is thus a superfluid, meaning it can flow without energy dissipation.

In a class of superconductors known as type II superconductors, including all known high-temperature superconductors, an extremely small amount of resistivity appears at temperatures not too far below the nominal superconducting transition when an electric current is applied in conjunction with a strong magnetic field, which may be caused by the electric current. This is due to the motion of magnetic vortices in the electronic superfluid, which dissipates some of the energy carried by the current. If the current is sufficiently small, the vortices are stationary, and the resistivity vanishes. The resistance due to this effect is tiny compared with that of non-superconducting materials, but must be taken into account in sensitive experiments. However, as the temperature decreases far enough below the nominal superconducting transition, these vortices can become frozen into a disordered but stationary phase known as a "vortex glass". Below this vortex glass transition temperature, the resistance of the material becomes truly zero.

## Superconducting Phase Transition

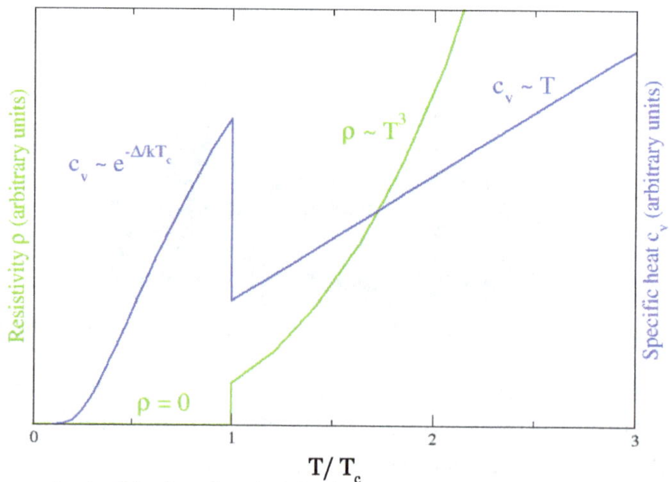


Behavior of heat capacity ($c_v$, blue) and resistivity (ρ, green) at the superconducting phase transition.

In superconducting materials, the characteristics of superconductivity appear when the temperature *T* is lowered below a critical temperature $T_c$. The value of this critical temperature varies from material to material. Conventional superconductors usually have critical temperatures ranging from around 20 K to less than 1 K. Solid mercury, for example, has a critical temperature of 4.2 K. As of 2009, the highest critical temperature found for a conventional superconductor is 39 K for magnesium diboride ($MgB_2$), although this material displays enough exotic properties that there is some doubt about classifying it as a "conventional" superconductor. Cuprate superconductors can have much higher critical temperatures: $YBa_2Cu_3O_7$, one of the first cuprate superconductors to be discovered, has a critical temperature of 92 K, and mercury-based cuprates have been found with critical temperatures in excess of 130 K. The explanation for these high critical temperatures remains unknown. Electron pairing due to phonon exchanges explains superconductivity in conventional superconductors, but it does not explain superconductivity in the newer superconductors that have a very high critical temperature.

Similarly, at a fixed temperature below the critical temperature, superconducting materials cease to superconduct when an external magnetic field is applied which is greater than the *critical magnetic field*. This is because the Gibbs free energy of the superconducting phase increases quadratically with the magnetic field while the free energy of the normal phase is roughly independent of the magnetic field. If the material superconducts in the absence of a field, then the superconducting phase free energy is lower than that of the normal phase and so for some finite value of the magnetic field (proportional to the square root of the difference of the free energies at zero magnetic field) the two free energies will be equal and a phase transition to the normal phase will occur. More generally, a higher temperature and a stronger magnetic field lead to a smaller fraction of the electrons in the superconducting band and consequently a longer London penetration depth of external magnetic fields and currents. The penetration depth becomes infinite at the phase transition.

The onset of superconductivity is accompanied by abrupt changes in various physical properties, which is the hallmark of a phase transition. For example, the electronic heat capacity is proportional to the temperature in the normal (non-superconducting) regime. At the superconducting transition, it suffers a discontinuous jump and thereafter ceases to be linear. At low temperatures, it varies instead as $e^{-\alpha/T}$ for some constant, α. This exponential behavior is one of the pieces of evidence for the existence of the energy gap.

The order of the superconducting phase transition was long a matter of debate. Experiments indicate that the transition is second-order, meaning there is no latent heat. However, in the presence of an external magnetic field there is latent heat, because the superconducting phase has a lower entropy below the critical temperature than the normal phase. It has been experimentally demonstrated that, as a consequence, when the magnetic field is increased beyond the critical field, the resulting phase transition leads to a decrease in the temperature of the superconducting material.

Calculations in the 1970s suggested that it may actually be weakly first-order due to the effect of long-range fluctuations in the electromagnetic field. In the 1980s it was shown theoretically with the help of a disorder field theory, in which the vortex lines of the superconductor play a major role, that the transition is of second order within the type II regime and of first order (i.e., latent heat) within the type I regime, and that the two regions are separated by a tricritical point. The results were strongly supported by Monte Carlo computer simulations.

## Meissner Effect

When a superconductor is placed in a weak external magnetic field H, and cooled below its transition temperature, the magnetic field is ejected. The Meissner effect does not cause the field to be completely ejected but instead the field penetrates the superconductor but only to a very small distance, characterized by a parameter λ, called the London penetration depth, decaying exponentially to zero within the bulk of the material. The Meissner effect is a defining characteristic of superconductivity. For most superconductors, the London penetration depth is on the order of 100 nm.

The Meissner effect is sometimes confused with the kind of diamagnetism one would expect in a perfect electrical conductor: according to Lenz's law, when a *changing* magnetic field is applied to a conductor, it will induce an electric current in the conductor that creates an opposing magnetic field. In a perfect conductor, an arbitrarily large current can be induced, and the resulting magnetic field exactly cancels the applied field.

The Meissner effect is distinct from this—it is the spontaneous expulsion which occurs during transition to superconductivity. Suppose we have a material in its normal state, containing a constant internal magnetic field. When the material is cooled below the critical temperature, we would observe the abrupt expulsion of the internal magnetic field, which we would not expect based on Lenz's law.

The Meissner effect was given a phenomenological explanation by the brothers Fritz and Heinz London, who showed that the electromagnetic free energy in a superconductor is minimized provided

$$\nabla^2 \boldsymbol{H} = \lambda^{-2} \boldsymbol{H}$$

where H is the magnetic field and λ is the London penetration depth.

This equation, which is known as the London equation, predicts that the magnetic field in a superconductor decays exponentially from whatever value it possesses at the surface.

A superconductor with little or no magnetic field within it is said to be in the Meissner state. The Meissner state breaks down when the applied magnetic field is too large. Superconductors can be divided into two classes according to how this breakdown occurs. In Type I superconductors, superconductivity is abruptly destroyed when the strength of the applied field rises above a critical value $H_c$. Depending on the geometry of the sample, one may obtain an intermediate state consisting of a baroque pattern of regions of normal material carrying a magnetic field mixed with regions of superconducting material containing no field. In Type II superconductors, raising the applied field past a critical value $H_{c1}$ leads to a mixed state (also known as the vortex state) in which an increasing amount of magnetic flux penetrates the material, but there remains no resistance to the flow of electric current as long as the current is not too large. At a second critical field strength $H_{c2}$, superconductivity is destroyed. The mixed state is actually caused by vortices in the electronic superfluid, sometimes called fluxons because the flux carried by these vortices is quantized. Most pure elemental superconductors, except niobium and carbon nanotubes, are Type I, while almost all impure and compound superconductors are Type II.

## London Moment

Conversely, a spinning superconductor generates a magnetic field, precisely aligned with the spin axis. The effect, the London moment, was put to good use in Gravity Probe B. This experiment measured the magnetic fields of four superconducting gyroscopes to determine their spin axes. This was critical to the experiment since it is one of the few ways to accurately determine the spin axis of an otherwise featureless sphere.

## History of Superconductivity

Superconductivity was discovered on April 8, 1911 by Heike Kamerlingh Onnes, who was studying the resistance of solid mercury at cryogenic temperatures using the recently produced liquid helium as a refrigerant. At the temperature of 4.2 K, he observed that the resistance abruptly disappeared. In the same experiment, he also observed the superfluid transition of helium at 2.2 K, without recognizing its significance. The precise date and circumstances of the discovery were only reconstructed a century later, when Onnes's notebook was found. In subsequent decades, superconductivity was observed in several other materials. In 1913, lead was found to superconduct at 7 K, and in 1941 niobium nitride was found to superconduct at 16 K.

Heike Kamerlingh Onnes (right), the discoverer of superconductivity. Paul Ehrenfest, Hendrik Lorentz, Niels Bohr stand to his left.

Great efforts have been devoted to finding out how and why superconductivity works; the important step occurred in 1933, when Meissner and Ochsenfeld discovered that superconductors expelled applied magnetic fields, a phenomenon which has come to be known as the Meissner effect. In 1935, Fritz and Heinz London showed that the Meissner effect was a consequence of the minimization of the electromagnetic free energy carried by superconducting current.

## London Theory

The first phenomenological theory of superconductivity was London theory. It was put forward by the brothers Fritz and Heinz London in 1935, shortly after the discovery that magnetic fields are expelled from superconductors. A major triumph of the equations of this theory is their ability to explain the Meissner effect, wherein a material exponentially expels all internal magnetic fields as it crosses the superconducting threshold. By using the London equation, one can obtain the dependence of the magnetic field inside the superconductor on the distance to the surface.

There are two London equations:

$$\frac{\partial \boldsymbol{j}_s}{\partial t} = \frac{n_s e^2}{m}\boldsymbol{E}, \qquad \nabla \times \boldsymbol{j}_s = -\frac{n_s e^2}{m}\boldsymbol{B}.$$

The first equation follows from Newton's second law for superconducting electrons.

## Conventional Theories (1950s)

During the 1950s, theoretical condensed matter physicists arrived at an understanding of "conventional" superconductivity, through a pair of remarkable and important theories: the phenomenological Ginzburg-Landau theory (1950) and the microscopic BCS theory (1957).

In 1950, the phenomenological Ginzburg-Landau theory of superconductivity was devised by Landau and Ginzburg. This theory, which combined Landau's theory of second-order phase transitions with a Schrödinger-like wave equation, had great success in explaining the macroscopic properties of superconductors. In particular, Abrikosov showed that Ginzburg-Landau theory predicts the division of superconductors into the two categories now referred to as Type I and Type II. Abrikosov and Ginzburg were awarded the 2003 Nobel Prize for their work (Landau had received the 1962 Nobel Prize for other work, and died in 1968). The four-dimensional extension of the Ginzburg-Landau theory, the Coleman-Weinberg model, is important in quantum field theory and cosmology.

Also in 1950, Maxwell and Reynolds *et al.* found that the critical temperature of a superconductor depends on the isotopic mass of the constituent element. This important discovery pointed to the electron-phonon interaction as the microscopic mechanism responsible for superconductivity.

The complete microscopic theory of superconductivity was finally proposed in 1957 by Bardeen, Cooper and Schrieffer. This BCS theory explained the superconducting current as a superfluid of Cooper pairs, pairs of electrons interacting through the exchange of phonons. For this work, the authors were awarded the Nobel Prize in 1972.

The BCS theory was set on a firmer footing in 1958, when N. N. Bogolyubov showed that the BCS wavefunction, which had originally been derived from a variational argument, could be obtained using a canonical transformation of the electronic Hamiltonian. In 1959, Lev Gor'kov showed that the BCS theory reduced to the Ginzburg-Landau theory close to the critical temperature.

Generalizations of BCS theory for conventional superconductors form the basis for understanding of the phenomenon of superfluidity, because they fall into the lambda transition universality class. The extent to which such generalizations can be applied to unconventional superconductors is still controversial.

## History

The first practical application of superconductivity was developed in 1954 with Dudley Allen Buck's invention of the cryotron. Two superconductors with greatly different values of critical magnetic field are combined to produce a fast, simple switch for computer elements.

Soon after discovering superconductivity in 1911, Kamerlingh Onnes attempted to make an electromagnet with superconducting windings but found that relatively low magnetic fields destroyed superconductivity in the materials he investigated. Much later, in 1955, G.B. Yntema succeeded in constructing a small 0.7-tesla iron-core electromagnet with superconducting niobium wire windings. Then, in 1961, J.E. Kunzler, E. Buehler, F.S.L. Hsu, and J.H. Wernick made the startling discovery that, at 4.2 kelvin, a compound consisting of three parts niobium and one part tin, was capable of supporting a current density of more than 100,000 amperes per square centimeter in a magnetic field of 8.8 tesla. Despite being brittle and difficult to fabricate, niobium-tin has since proved extremely useful in supermagnets generating magnetic fields as high as 20 tesla. In 1962 T.G. Berlincourt and R.R. Hake discovered that alloys of niobium and titanium are suitable for applications up to 10 tesla. Promptly thereafter, commercial production of niobium-titanium supermagnet wire commenced at Westinghouse Electric Corporation and at Wah Chang Corporation. Although niobium-titanium boasts less-impressive superconducting properties than those of niobium-tin, niobium-titanium has, nevertheless, become the most widely used "workhorse" supermagnet material, in large measure a consequence of its very-high ductility and ease of fabrication. However, both niobium-tin and niobium-titanium find wide application in MRI medical imagers, bending and focusing magnets for enormous high-energy-particle accelerators, and a host of other applications. Conectus, a European superconductivity consortium, estimated that in 2014, global economic activity for which superconductivity was indispensable amounted to about five billion euros, with MRI systems accounting for about 80% of that total.

In 1962, Josephson made the important theoretical prediction that a supercurrent can flow between two pieces of superconductor separated by a thin layer of insulator. This phenomenon, now called the Josephson effect, is exploited by superconducting devices such as SQUIDs. It is used in the most accurate available measurements of the magnetic flux quantum $\Phi_0 = h/(2e)$, where $h$ is the Planck constant. Coupled with the quantum Hall resistivity, this leads to a precise measurement of the Planck constant. Josephson was awarded the Nobel Prize for this work in 1973.

In 2008, it was proposed that the same mechanism that produces superconductivity could produce a superinsulator state in some materials, with almost infinite electrical resistance.

## High-temperature Superconductivity

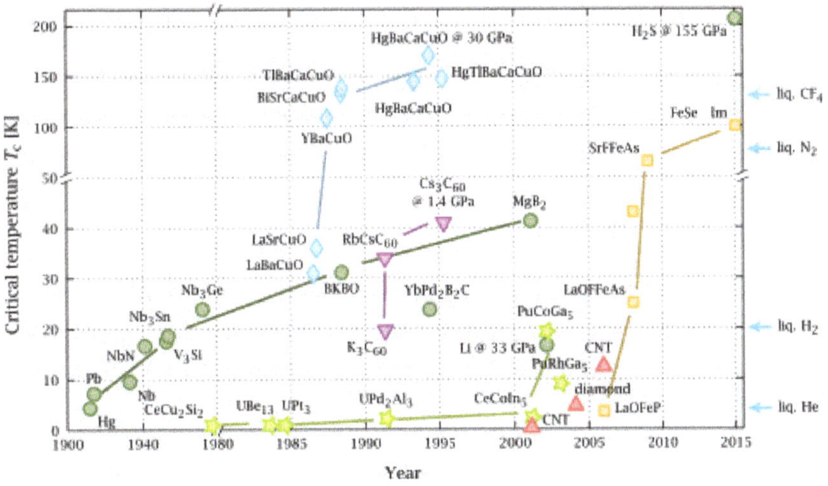


Timeline of superconducting materials.

Until 1986, physicists had believed that BCS theory forbade superconductivity at temperatures above about 30 K. In that year, Bednorz and Müller discovered superconductivity in a lanthanum-based cuprate perovskite material, which had a transition temperature of 35 K (Nobel Prize in Physics, 1987). It was soon found that replacing the lanthanum with yttrium (i.e., making YBCO) raised the critical temperature to 92 K.

This temperature jump is particularly significant, since it allows liquid nitrogen as a refrigerant, replacing liquid helium. This can be important commercially because liquid nitrogen can be produced relatively cheaply, even on-site. Also, the higher temperatures help avoid some of the problems that arise at liquid helium temperatures, such as the formation of plugs of frozen air that can block cryogenic lines and cause unanticipated and potentially hazardous pressure buildup.

Many other cuprate superconductors have since been discovered, and the theory of superconductivity in these materials is one of the major outstanding challenges of theoretical condensed matter physics. There are currently two main hypotheses – the resonating-valence-bond theory, and spin fluctuation which has the most support in the research community. The second hypothesis proposed that electron pairing in high-temperature superconductors is mediated by short-range spin waves known as paramagnons.

Since about 1993, the highest-temperature superconductor has been a ceramic material consisting of mercury, barium, calcium, copper and oxygen ($HgBa_2Ca_2Cu_3O_{8+\delta}$) with $T_c = 133–138$ K. The latter experiment (138 K) still awaits experimental confirmation, however.

In February 2008, an iron-based family of high-temperature superconductors was discovered. Hideo Hosono, of the Tokyo Institute of Technology, and colleagues found lanthanum oxygen fluorine iron arsenide ($LaO_{1-x}F_xFeAs$), an oxypnictide that superconducts below 26 K. Replacing the lanthanum in $LaO_{1-x}F_xFeAs$ with samarium leads to superconductors that work at 55 K.

In May 2014, hydrogen sulfide ($H_2S$) was predicted to be a high-temperature superconductor with a transition temperature of 80 K at 160 gigapascals of pressure. In 2015, $H_2S$ has been observed to exhibit superconductivity at below 203 K but at extremely high pressures — around 150 gigapascals.

## Applications

Superconducting magnets are some of the most powerful electromagnets known. They are used in MRI/NMR machines, mass spectrometers, the beam-steering magnets used in particle accelerators and plasma confining magnets in some tokamaks. They can also be used for magnetic separation, where weakly magnetic particles are extracted from a background of less or non-magnetic particles, as in the pigment industries.

In the 1950s and 1960s, superconductors were used to build experimental digital computers using cryotron switches. More recently, superconductors have been used to make digital circuits based on rapid single flux quantum technology and RF and microwave filters for mobile phone base stations.

Superconductors are used to build Josephson junctions which are the building blocks of SQUIDs (superconducting quantum interference devices), the most sensitive magnetometers known. SQUIDs are used in scanning SQUID microscopes and magnetoencephalography. Series of Josephson devices are used to realize the SI volt. Depending on the particular mode of operation, a superconductor-insulator-superconductor Josephson junction can be used as a photon detector or as a mixer. The large resistance change at the transition from the normal- to the superconducting state is used to build thermometers in cryogenic micro-calorimeter photon detectors. The same effect is used in ultrasensitive bolometers made from superconducting materials.

Other early markets are arising where the relative efficiency, size and weight advantages of devices based on high-temperature superconductivity outweigh the additional costs involved. For example, in wind turbines the lower weight and volume of superconducting generators could lead to savings in construction and tower costs, offsetting the higher costs for the generator and lowering the total LCOE.

Promising future applications include high-performance smart grid, electric power transmission, transformers, power storage devices, electric motors (e.g. for vehicle propulsion, as in vactrains or maglev trains), magnetic levitation devices, fault current limiters, enhancing spintronic devices with superconducting materials, and superconducting magnetic refrigeration. However, superconductivity is sensitive to moving magnetic fields so applications that use alternating current (e.g. transformers) will be more difficult to develop than those that rely upon direct current. Compared to traditional power lines superconducting transmission lines are more efficient and require only a fraction of the space, which would not only lead to a better environmental performance but could also improve public acceptance for expansion of the electric grid.

## Nobel Prizes for Superconductivity

- Heike Kamerlingh Onnes (1913), "for his investigations on the properties of matter at low temperatures which led, inter alia, to the production of liquid helium"
- John Bardeen, Leon N. Cooper, and J. Robert Schrieffer (1972), "for their jointly developed theory of superconductivity, usually called the BCS-theory"
- Leo Esaki, Ivar Giaever, and Brian D. Josephson (1973), "for their experimental discoveries regarding tunneling phenomena in semiconductors and superconductors, respectively," and "for his theoretical predictions of the properties of a supercurrent through a tunnel barrier, in particular those phenomena which are generally known as the Josephson effects"
- Georg Bednorz and K. Alex Müller (1987), "for their important break-through in the discovery of superconductivity in ceramic materials"
- Alexei A. Abrikosov, Vitaly L. Ginzburg, and Anthony J. Leggett (2003), "for pioneering contributions to the theory of superconductors and superfluids"
- Michael Kosterlitz, Duncan Haldane, David J. Thouless (2016)

## Historical Overview

Superconductivity is an exciting area of physics where truly exotic phenomena can now be demonstrated with relatively inexpensive instruments/equipment. One of the most appealing demnstrations is that of a superconductor moving on magnetic rails in a levitated fashion thereby suggesting futuristic applications.

A futuristic concept of a magnetically levitated train using permanent magnets for the rails and a superconductor for the carriage is demonstrated here. The superconductor floats on the rails thereby eliminating friction.

The history of superconductivity is intimately connected with the attempts to obtain lower and lower temperatures. This, in turn, was driven by the research on properties of gases as a function of temperature.

Various scales have been used to record the temperature. The Fahrenheit scale dates back to 1724. Daniel Gabriel Fahrenheit's original scale used the temperature under the human armpit as one of its three fixed points. Anders Celsius' original scale had two fixed points, the boiling point 0 C and freezing point of water 100 C. This is exactly inverse of what we have currently. This correction was done by Swedish botanist and taxonomist Carl Linnaeus.

The motivation behind this was to test the equation-of-state for non-ideal gases. Naturally, in the beginning, it made sense to study simple diatomic gases such as nitrogen, hydrogen, oxygen, helium, and argon. These gases condense only at very low temperatures and hence one could study them in a large temperature range. Somewhat driven by the above objectives, researchers developed techniques to cool gases to low temperatures and thereby also to eventually condense them.

Significant work was done in the early nineteenth century by Michael Faraday at the Royal Institution in London. Faraday was an apprentice bookbinder from a poor family. He managed to secure a job with the Royal Institution as a scientific assistant with the eminent chemist Sir Humphrey Davy, mainly on the strength of presenting him with a bound version of notes he had taken at some of Davy's public lectures. Davy's wife still treated him like a servant. He accidentally discovered a way to make liquid chlorine.

In 1811, Davy had showed that the crystals obtained by passing chlorine gas through a nearly freezing dilute solution of calcium chloride were a compound of chlorine and water; chlorine hydrate ($Cl_2H_2O$). At Davy's suggestion, Faraday performed some experiments. He heated a sealed glass tube containing the above crystals. The other end of the tube was submerged in ice. An oily liquid condensed at the cold end which, experiments showed, was chlorine. The high pressure in the tube allowed the liquefaction to take place at a lower temperature (-34 C) than it would at ambient pressure. Davy was the first to liquefy an element though compounds had been liquefied earlier. This technique was then used to liquefy other gases like ammonia, hydrogen sulfide, nitrogen dioxide etc. Faraday called hydrogen, nitrogen, and oxygen permanent gases since he was unable to liquefy them. Actually, it was the limitation on obtaining sufficient pressure which prevented him from liquefying the other gases. Another original procedure for cooling and liquefying gases was discovered by accident. Louis Paul Cailletet (son of a metallurgist) had set up a lab in his father's foundry. He was trying to liquefy acetylene. During the pressurisation, his tube sprang a leak. As the gas escaped through the leak, a faint mist was formed near the outlet which quickly disappeared. The first instinct was to suspect water impurity in the starting material. But this also happened in pure samples. Another way to obtain lower temperatures was therefore to release the pressure instead of augmenting it. Thus was born a new scheme to liquefy gases. He then carried out the process (reported in the academy of science in Paris in 1877) for oxygen successfully. He further went on to liquefy nitrogen and carbon monoxide. Another method was developed by swiss chemist Raoul-Peirre Pictet in Geneva based on a cascade process.

This involved the use of cascade-like processes where the least volatile gas is cooled and condensed first. Its evaporating vapour is then used to cool the more volatile gas and so on. As a result various gases were liquefied.

Table : The boiling points of various common gases at atmospheric pressure.

| Gas | Boiling point (K) |
|---|---|
| Oxygen | 90.1 |
| Argon | 87.1 |
| Nitrogen | 77.3 |
| Hydrogen | 20.3 |
| Helium | 4.2 |

where the boiling points of various gases are given. In 1898, James Dewar (best known for his invention of the double walled glass container with vacuum in the intermediate space, for storing cryogenic liquidsin UK, succeeded in liquefying hydrogen at about 20K (To this day, the containers used in low-temperature research labs are called dewars).

The figure shows dewars for storing/transporting cryogenic liquids. The containers are double walled with the space between them evacuated to a high degree. This minimises the loss of the liquid in the container due to evaporation.

Cooling by rapid expansion had actually been established earlier in 1852 by James Prescott Joule and William Thompson (later Lord Kelvin) as the Joule-Thompson effect. Using this effect in a closed cycle mode, Carl Paul Gottfried von Linde patented and installed in Munich a commercial refrigeration system in 1873. Now, apparently, only hydrogen was left to liquefy. William Ramsay, in University College London, isolated helium (earlier discovered spectroscopically in the Sun) while looking for Argon. He got the Nobel prize in 1904 for his discovery of He, Ne, Ar, Kr, Xe. In 1877, Sir James Dewar became Fullerian Professor of chemistry (a chair earlier held by Faraday) at the Royal Institution. He obtained a Cailletet apparatus and started his work liquefying gases. produced solid oxygen in 1886. He wanted his work/demo to be visible to the public while giving a talk etc. therefore wanted to have transparent vessels but then there is a problem of thermal isolation. He invented the double walled dewar for this purpose. On 10 May 1898 Dewar produced 20 cc of liq H2. He deduced a temp of 20.28 K, eventually even solidified H2 at 14 K. helim still remained. Dewar and Ramsay would have been a great pair to liquefy helium but they had a falling out. Ramsay announced to the Royal Society that in fact Dewar was not the first to liq H but that it had already been done by Prof. Olszewski from Poland. It turned out that he was wrong and even Olszewski admitted that Dewar was first.

Subsequently, great efforts were made to liquefy the only remaining gas i.e., helium. Using iterative Joule-Thomson cooling, Heike Kammerlingh Onnes succeeded in liquefying helium gas at the University of Leiden in the Netherlands on July 10, 1908.

Finally, it was Onnes in Leiden who succeeded in liquefying helium using a cascade process. First he liquefied hydrogen but only in 1906. His brother was the director of the office of commercial intelligence in Amsterdam and was able to source large quantities of monazite sands from North Carolina. After a period of three years, he had 300 litres of helium gas at his disposal. in 1908, production rate was a litre of liquid every 3-4 hrs and a decade later, about 2 litres per hour.

This can be considered as the beginning of research on superconductivity and more generally, low-temperature physics. In the following years, Kammerlingh Onnes started investigating the electrical resistance of metals as a function of temperature. Mercury is a liquid at room temperature and distillation techniques could be used to purify it. With the idea of examining properties of a high-purity metal, he took up the measurement of electrical resistivity of mercury in early 1911. Mercury was filled in a U-shaped tube with wires inserted at both ends and its resistance was measured as a function of temperature. The result obtained was quite remarkable in that the resistance decreased with decreasing temperature and then dropped precipitously to zero at about 4.2 K.

This was the first ever observation of superconductivity (called supraconductivity by Onnes at that time). He was deciding between gold and mercury for his first measurement and it is fortunate that he chose mercury since gold is not superconducting. Onnes realized the commercial potential of his discovery of superconductivity and began examining other metals. Tin and lead were the next elements to be found superconducting by him. Since the Leiden lab. had a near monopoly in the production of liquid helium, Onnes and his co-workers were the leaders in low-temperature physics for many years. Subsequent to the original discovery of Onnes of zero resistance in Hg at 4.2 K, many new superconductors and allied phenomena were discovered. In one of his experiments Onnes started a current in a loop of lead wire cooled to 4 K. Even after a year the current was still flowing without any noticeable change. This was called a persistent current by Onnes. Kammerlingh Onnes was awarded the Nobel prize in 1913. The citation read that the prize was for his investigations on the properties of matter at low temperatures which led, inter alia, to the production of liquid helium, though not really for the discovery of superconductivity. Finally Onnes was the person who also found that superconductivity could be destroyed by an applied magnetic field called the critical field. The empirical relation for the temperature variation of the critical field was found to be $H_c(T) = H_c(0)[1-(\frac{T}{T_c})^2]$. Here, $T_c$ (the critical temperature) is the temperature below which the substance is superconducting when no magnetic field is applied and $H_c(T)$ is the magnetic field necessary to decrease the critical temperature to $T$.

A theoretical explanation for any of the observations related to superconductivity was lacking at that time though people believed that it lay in quantum mechanics. Researchers kept working to discover new superconductors and to learn new properties in order to shed more light on this new and astonishing phenomenon. The next important finding was made by Meissner and Ochsenfeld who, in 1933, found that superconductors are not merely perfect conductors (i.e., having a zero resistance) but also exclude a magnetic field from their interior.

Following the research work geared towards the exploration of possible superconductivity in elements, attention was focused on alloys.

In Table below, the critical temperatures of various elements and compounds along with the year in which they were discovered is given.

Table : The $T_c$ of various compounds is given along with their year of discovery.

| Chemical formula | $T_c$ (K) | year of discovery |
|---|---|---|
| Hg | 4.15 | 1911 |

| | | |
|---|---|---|
| V | 5.38 | 1934 |
| Pb | 7.19 | 1922 |
| Nb | 9.5 | 1930 |
| $LiTi_2O_4$ | 12 | 1973 |
| NbN | 16 | 1941 |
| $V_3Ga$ | 16.5 | 1960 |
| $V_3Si$ | 17.5 | 1953 |
| $Nb_3Al$ | 17.5 | 1969 |
| $Nb_3Sn$ | 18 | 1952 |
| $Nb_3Ge$ | 23.2 | 1973 |

Experimental work continued to elucidate various aspects of superconductivity. On the theoretical front concerted efforts were on to provide an explanation. There were early efforts by brothers London to explain flux expulsion from a superconductor (London model ca. 1920).

Fritz London obtained obtained appointments in theoretical Physics with Paul Ewald in Stuttgart and then with Arnold Sommerfeld in Munich (six of his students were awarded the Nobel prize.). With Walter Heitler, he advanced the Heitler-London theory for hydrogen molecule. His younger brother Heinz London studied under Franz Simon, who was an expert at low-temperature techniques, at Breslau. He (Heinz London) later changed his area of research from experiment to theory. These (early 30s) were turbulent times in Europe due to the spread of fascism. Frederick Lindemann was at Oxford University (UK) and was trying to help the Jewish scientists by enabling them to escape from Germany. His objectives behind this were two-fold. In those days the reputation of the Physics Department at Oxford was much lower compared to the Cambridge Cavendish lab. So by getting eminent scientists, he could enhance the reputation of the lab as also help the Jewish people. He managed to get Franz Simon, F. London, Kurt Mendelssohn, Nicholas Kurti there. In addition, he also invited Erwin Schroedinger and A. Einstein. Einstein did not stay for long and quickly went away to Princeton University in the USA. London coined the term "macroscopic quantum phenomena" for superconductivity.

While, it might have been an ad-hoc phenomenological model, it did introduce an important physical length scale, namely the penetration depth, associated with superconductors. The microscopic theory of superconductivity was proposed several decades later in 1957 by J. Bardeen (then a Professor of Physics at University of Illinois Urbana Champagne, USA as also a Nobel prize winner for the invention of the point contact transistor) along with his post-doctoral fellow L. Cooper and his PhD student J.R. Schrieffer.

Bardeen, Brattain and Shockley got the Nobel prize for inventing the point contact transistor. It is said that most of the work was done by the former two while Shockley (who was the boss) claimed the credit for it. Interestingly, after their work on the point contact transistor, Shockley forbade them to work on the junction transistor on which he worked all by himself and did not inform the other two. Bardeen later quit Bell labs and joined University of Illinois at Urbana Champaigne. Shockley also quit Bell Labs and formed his own company which eventually failed. Bardeen got his PhD with Eugene Wigner at Princeton and was interested in superconductivity since then. He first worked on the problem while in Minnesota in 1941. Froehlich proposed a Hamiltonian relevant to the problem of superconductivity following the discovery of isotope effect.

Many of the already observed aspects were explained by the theory. Further, predictions were made which were verified in the years that followed. Bardeen, Cooper, and Schrieffer were awarded the Nobel prize for the theory of superconductivity (also called the BCS theory) in 1972. Even before the BCS theory, a general phenomenology of phase transitions was given by the Ginzburg-Landau theory in 1950. They expressed the free energy as an in powers of an order parameter which characterized the superconducting state. Using minimisation procedures, they obtained further insight into the initially unknown parameters that were included in the theory. With this they obtained a handle on various properties of a superconductor. On the applications front, superconductors ($Nb_3Sn$ and $Nb_3Ti$) were already being used commercially for solenoid magnets, SQUID sensors, etc.

As a consequence of intense research, new superconducting materials were being discovered and the $T_c$ was rising at about 2.8 K/decade. Superconductivity has been found in a diverse set of materials including organic and heavy-Fermion materials. However, it was the discovery of superconductivity in Ba doped $La_2CuO_4$ ($T_c$ 30K) by Bednorz and Mueller in 1986 which revolutionized research in this area. While the value of $T_c$ observed by them was not strictly out of the range predicted by the BCS theory, it overthrew the existing beliefs which (empirically) suggested that superconductors should have high crystallographic symmetry such as cubic (the cuprate superconductors are strongly anisotropic and layered in structure) or that they may follow a universal $T_c$ vs. average valence electron count curve. Soon after the initial discovery by Bednorz and Mueller (which won them the Physics Nobel prize in 1987), extensive work by scientists led to the discovery of families of cuprate superconductors with the highest onset- $T_c$ in the region of 150 K.

Bednorz and Mueller were awarded the Nobel prize for discovering high-Tc superconductivity in the cuprates. There is an interesting story about how their paper got published. Their work was performed at the IBM research labs in Rueschlikon near Zurich. An approval from their boss was needed to send the paper for publication. Binnig was their boss and he had just been informed that he, along with Rohrer, had won the Nobel prize in Physics for the invention of the scanning tunneling microscope STM. In this party-like atmosphere in the lab., Bednorz and Mueller asked for his autograph on a sheet of paper which was actually the approval for sending their paper for publication. They sent the paper to Z. Phys. and requested the editor to publish without refereeing which was agreed to. Following the research of Bednorz and Mueller, people worked feverishly to find superconductors with a even higher Tc. Paul Chu, after discovering superconductivity at 90 K in $YBa_2Cu_3O_7$, sent the manuscript to Phys. Rev. Lett. for publication but first requested the editor to publish it without any refereeing which was not agreed to. He then requested that the manuscript be sent to his suggested list of referees, which was agreed to by the editor. In the initial version of the manuscript, he wrote Yb instead of Y in the chemical formula and also gave a different stoichiometry. This was evidently to prevent the paper from falling into other hands. At the time of proof corrections, he changed the formula to the correct one. In spite of this precaution, some people published papers saying that they had discovered superconductivity with Yb in the formula. This seems to indicate that information leaked from referees. It is another matter that even the Yb-based compound is superconducting. Also, in the early days of high-Tc, people were not able to make single-phase samples and hence not knowing the correct stoichiometry did not hurt so much.

The cuprate superconductors are unconventional in many ways and efforts continue to-date to develop a theoretical framework which will explain various features of the high-$T_c$ superconductors. In the following years superconductivity was also discovered in fullerenes ($M_3C_{60}$ where M = Na, K, Rb, Cs). In some ways, this lead to the now independent field of nanotubes. Another notable discovery of superconductivity was in $MgB_2$ ($T_c$ 40K) in 2001. The unusual thing about it was that $MgB_2$ had been even commercial available for a long time before it was investigated and found to be superconducting. Experimental and theoretical work indicates that superconductivity in $MgB_2$ is driven by the conventional electron-phonon interaction. The most recent advance in superconductivity research is the discovery of pnictide and chalcogenide families ($REFeAsO_{1-x}F_x$, $REFe_2As_2$, FeSe, etc. where RE is a rare earth element) where the highest $T_c$ is reported to be around 56 K. In summary, the last two decades have witnessed a tremendous advancement in superconductivity research and the future promises to be full of surprises.

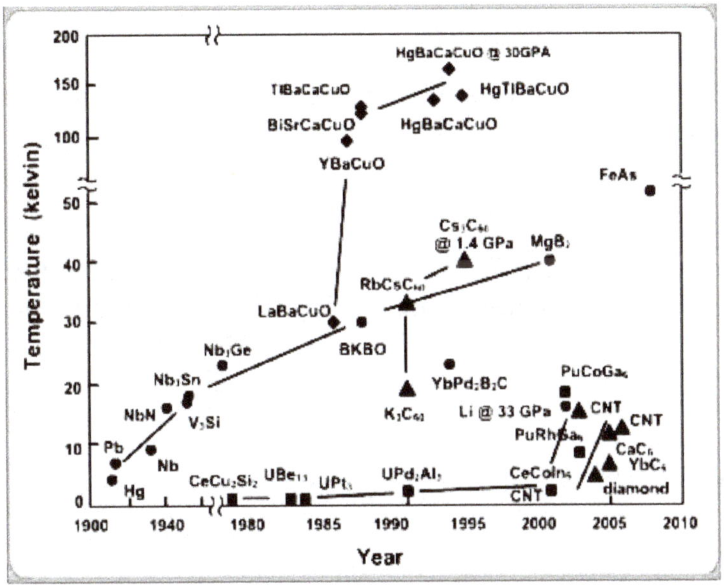


As a consequence of research over the decades, new materials have been discovered with higher and higher superconducting transition temperatures. A steep rise has taken place with the discovery of cuprate superconductors. This historical development is traced in the figure.

## Superconducting Wire

Superconducting wire is wire made of superconductors. When cooled below its transition temperature, it has zero electrical resistance. Most commonly, conventional superconductors such as niobium-titanium are used, but high-Temperature superconductors such as YBCO are entering the market. Superconducting wire's advantages over copper or aluminum include higher maximum current densities and zero power dissipation. Its disadvantages include the cost of refrigeration of the wires to superconducting temperatures (often requiring cryogens such as liquid helium or liquid nitrogen), the danger of the wire quenching (a sudden loss of superconductivity), the inferior mechanical properties of some superconductors, and the cost of wire materials and construction. Its main application is in superconducting magnets, which are used in scientific and medical equipment where high magnetic fields are necessary.

An example of a wire ($V_3Ga$ alloy) used in a superconducting magnet.

## Important Parameters of SC Wires/Tapes/Conductors

The construction and operating temperature will typically be chosen to maximise:

- critical temperature $T_c$, below this temperature the wire becomes a superconductor
- critical current density $J_c$, maximum super-current a superconducting wire can carry per unit cross sectional area.

## LTS Wire

Low-temperature superconductor (LTS) wires are made from superconductors with low critical temperature, such as $Nb_3Sn$(niobium-tin) and NbTi(niobium-titanium). Often the superconductor is in filament form in a copper or aluminium matrix which carries the current should the superconductor quench for any reason. The superconductor filaments can form a third of the total volume of the wire.

## Preparation

## Wire Drawing

The normal wire drawing process can be used for malleable alloys such as niobium-titanium.

## Surface Diffusion

Vanadium-gallium ($V_3$) can be prepared by surface diffusion where the high temperature component as a solid is bathed in the other element as liquid or gas. When all components remain in the solid state during high temperature diffusion this is known as the bronze process.

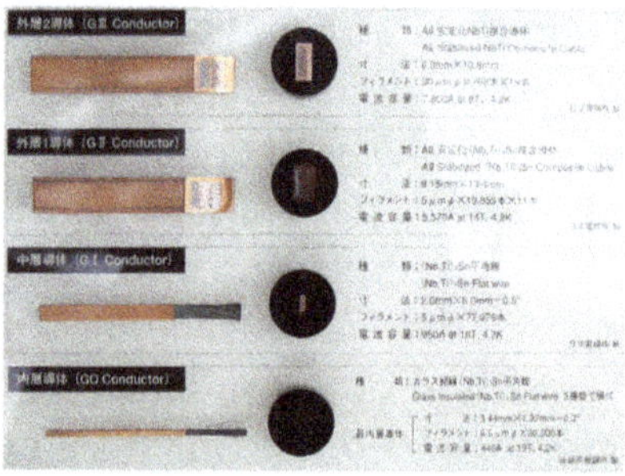

Cross sections of various $(Nb,Ti)_3Sn$ composite superconducting cables and wires. (440 to 7,800 A in 8 to 19 Tesla fields).

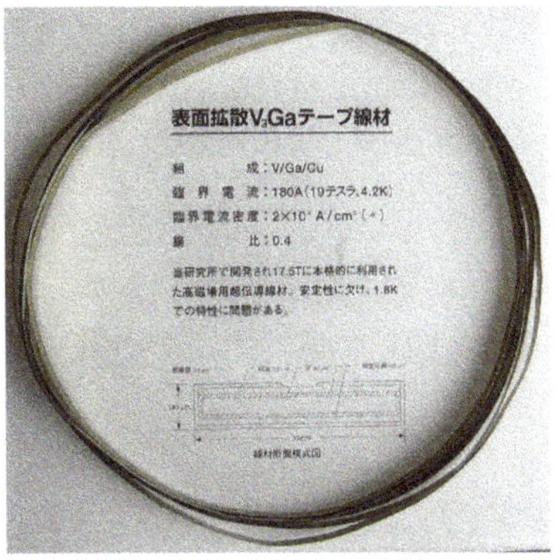


$V_3Ga$ superconducting tape (10×0.14 mm cross section). A vanadium core is covered with 15 μm $V_3Ga$ layer, then 20 μm bronze (stabilizing layer) and 15 μm insulating layer. Critical current 180 A (19.2 tesla, 4.2 K), critical current density 20 kA/cm².

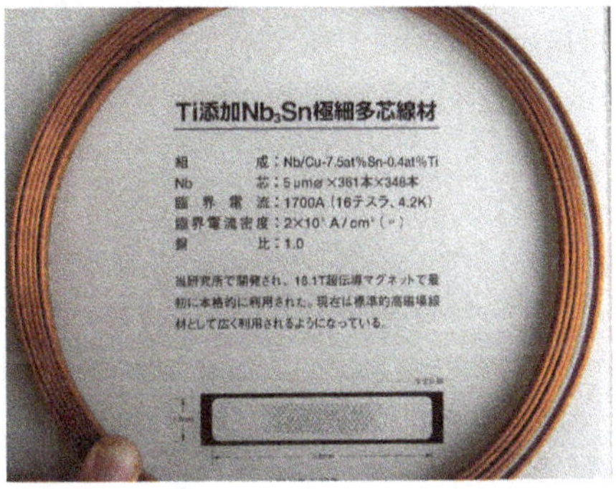


Nb/Cu-7.5at%Sn-0.4at%Ti tape (9.5×1.8 mm cross section) originally developed for an 18.1 T magnet. Nb core: 361×348 packs of 5 μm dia. filaments. Critical current 1700 A (16 tesla, 4.2 K), critical current density 20 kA/cm².

## HTS Wire

High-temperature superconductor (HTS) wires are made from superconductors with high critical temperature (high-temperature superconductivity), such as YBCO and BSCCO.

## Powder-in-tube

The powder-in-tube (PIT, or *oxide powder in tube*, OPIT) process is often used for making electrical conductors from brittle superconducting materials such as niobium-tin or magnesium diboride, and ceramic cuprate superconductors such as BSCCO. It has been used to form wires of the iron pnictides. (PIT is not used for yttrium barium copper oxide as it does not have the weak layers required to generate adequate 'texture' (alignment) in the PIT process.)

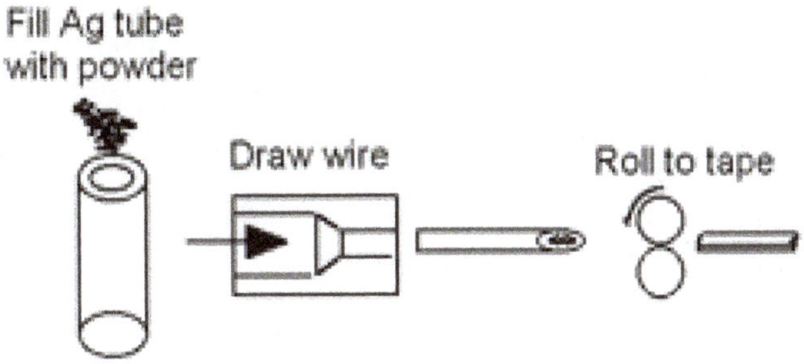


Simplified diagram of the PIT process.

This process is used because the high-temperature superconductors are too brittle for normal wire forming processes. The tubes are metal, often silver. Often the tubes are heated to react the mix of powders. Once reacted the tubes are sometimes flattened to form a tape-like conductor. The resulting wire is not as flexible as conventional metal wire, but is sufficient for many applications.

There are *in situ* and *ex situ* variants of the process, as well a 'double core' method that combines both.

## Coated Superconductor Tape or Wire

The coated superconductor tapes are known as second generation superconductor wires. These wires are in a form of a metal tape of about 10 mm width and about 100 micrometer thickness, coated with superconductor materials such as YBCO. A few years after the discovery of High-temperature superconductivity materials such as the YBCO, it was demonstrated that epitaxial YBCO thin films grown on lattice matched single crystals such as magnesium oxide MgO, strontium titanate ($SrTiO_3$) and sapphire had high supercritical current densities of 1–4 $MA/cm^2$. However, a lattice-matched flexible material was needed for producing a long tape. YBCO films deposited directly on metal substrate materials exhibit poor superconducting properties. It was demonstrated that a c-axis oriented yttria-stabilized zirconia (YSZ) intermediate layer on a metal substrate can yield YBCO films of higher quality, which had still one to two orders less critical current density than that produced on the single crystal substrates.

The breakthrough came with the invention of ion beam-assisted deposition (IBAD) technique to produce of biaxially aligned yttria-stabilized zirconia (YSZ) thin films on metal tapes.

The biaxial YSZ film acted as a lattice matched buffer layer for the epitaxial growth of the YBCO films on it. These YBCO films achieved critical current density of more than 1 $MA/cm^2$. Other buffer layers such as cerium oxide ($CeO_2$ and magnesium oxide (MgO) were produced using the IBAD technique for the superconductor films.

Smooth substrates with roughness in the order of 1 nm are essential for the high quality superconductor films. Initially hastelloy substrates were electro polished to create a smoothed surface. Hastelloy is a nickel based alloy capable of withstanding temperatures up to 800C without melting or heavily oxidizing. Currently a coating technique known as "spin on glass" or " solution deposition planarization" is used to smooth the substrate surface.

Recently YBCO coated superconductor tapes capable of carrying more than 500 $A/cm^2$ at 77 K and 1000 $A/cm^2$ at 30 K under high magnetic field have been demonstrated.

## Chemical Vapor Deposition

CVD is used for YBCO coated tapes.

## Hybrid Physical-chemical Vapor Deposition

HPCVD can be used for thin-film magnesium diboride. (Bulk $MgB_2$ can be made by PIT or reactive Mg liquid infiltration.)

# Color Superconductivity

Color superconductivity is a phenomenon predicted to occur in quark matter if the baryon density is sufficiently high (well above nuclear density) and the temperature is not too high (well below $10^{12}$ kelvin). Color superconducting phases are to be contrasted with the normal phase of quark matter, which is just a weakly interacting Fermi liquid of quarks.

In theoretical terms, a color superconducting phase is a state in which the quarks near the Fermi surface become correlated in Cooper pairs, which condense. In phenomenological terms, a color superconducting phase breaks some of the symmetries of the underlying theory, and has a very different spectrum of excitations and very different transport properties from the normal phase.

## Description

### Analogy with Superconducting Metals

It is well known that at low temperature many metals become superconductors. A metal can be viewed as a Fermi liquid of electrons, and below a critical temperature, an attractive phonon-mediated interaction between the electrons near the Fermi surface causes them to pair up and form a condensate of Cooper pairs, which via the Anderson-Higgs mechanism makes the photon massive, leading to the characteristic behaviors of a superconductor; infinite conductivity and the exclusion of magnetic fields (Meissner effect). The crucial ingredients for this to occur are:

1. a liquid of charged fermions.
2. an attractive interaction between the fermions
3. low temperature (below the critical temperature)

These ingredients are also present in sufficiently dense quark matter, leading physicists to expect that something similar will happen in that context:

1. quarks carry both electric charge and color charge;
2. the strong interaction between two quarks is powerfully attractive;
3. the critical temperature is expected to be given by the QCD scale, which is of order 100 MeV, or $10^{12}$ kelvin, the temperature of the universe a few minutes after the big bang, so quark matter that we may currently observe in compact stars or other natural settings will be below this temperature.

The fact that a Cooper pair of quarks carries a net color charge, as well as a net electric charge, means that some of the gluons (which mediate the strong interaction just as photons mediate electromagnetism) become massive in a phase with a condensate of quark Cooper pairs, so such a phase is called a "color superconductor". Actually, in many color superconducting phases the photon itself does not become massive, but mixes with one of the gluons to yield a new massless "rotated photon". This is an MeV-scale echo of the mixing of the hypercharge and $W_3$ bosons that originally yielded the photon at the TeV scale of electroweak symmetry breaking.

## Diversity of Color Superconducting Phases

Unlike an electrical superconductor, color-superconducting quark matter comes in many varieties, each of which is a separate phase of matter. This is because quarks, unlike electrons, come in many species. There are three different colors (red, green, blue) and in the core of a compact star we expect three different flavors (up, down, strange), making nine species in all. Thus in forming the Cooper pairs there is a $9\times9$ color-flavor matrix of possible pairing patterns. The differences between these patterns are very physically significant: different patterns break different symmetries of the underlying theory, leading to different excitation spectra and different transport properties.

It is very hard to predict which pairing patterns will be favored in nature. In principle this question could be decided by a QCD calculation, since QCD is the theory that fully describes the strong interaction. In the limit of infinite density, where the strong interaction becomes weak because of asymptotic freedom, controlled calculations can be performed, and it is known that the favored phase in three-flavor quark matter is the *color-flavor-locked* phase. But at the densities that exist in nature these calculations are unreliable, and the only known alternative is the brute-force computational approach of lattice QCD, which unfortunately has a technical difficulty (the "sign problem") that renders it useless for calculations at high quark density and low temperature.

Physicists are currently pursuing the following lines of research on color superconductivity:

- Performing calculations in the infinite density limit, to get some idea of the behavior at one edge of the phase diagram.
- Performing calculations of the phase structure down to medium density using a highly simplified model of QCD, the Nambu-Jona-Lasinio (NJL) model, which is not a controlled approximation, but is expected to yield semi-quantitative insights.
- Writing down an effective theory for the excitations of a given phase, and using it to calculate the physical properties of that phase.
- Performing astrophysical calculations, using NJL models or effective theories, to see if there are observable signatures by which one could confirm or rule out the presence of specific color superconducting phases in nature (i.e. in compact stars).

## Possible Occurrence in Nature

The only known place in the universe where the baryon density might possibly be high enough to produce quark matter, and the temperature is low enough for color superconductivity to occur, is the core of a compact star (often called a "neutron star", a term which prejudges the question of its actual makeup). There are many open questions here:

- We do not know the critical density at which there would be a phase transition from nuclear matter to some form of quark matter, so we do not know whether compact stars have quark matter cores or not.
- On the other extreme, it is conceivable that nuclear matter in bulk is actually metastable, and decays into quark matter (the "stable strange matter hypothesis"). In this case, compact stars would consist completely of quark matter all the way to their surface.

- Assuming that compact stars do contain quark matter, we do not know whether that quark matter is in a color superconducting phase or not. At infinite density one expects color superconductivity, and the attractive nature of the dominant strong quark-quark interaction leads one to expect that it will survive down to lower densities, but there may be a transition to some strongly coupled phase (e.g. a Bose–Einstein condensate of spatially bound di- or hexaquarks).

## History

The first physicists to realize that Cooper pairing could occur in quark matter were D. D. Ivanenko and D. F. Kurdgelaidze of Moscow State University, in 1969. However, their insight was not pursued until the development of QCD as the theory of the strong interaction in the early 1970s. In 1977 Stephen Frautschi, a professor at Caltech, and his graduate student Bertrand Barrois realized that QCD predicts Cooper instability leading to a colorless 6-quark Bose–Einstein condensate in high density quark matter, and coined the term "color superconductivity". Barrois was able to get part of his work published in the journal *Nuclear Physics*, but that journal rejected the longer manuscript based on his thesis, which anticipated later results such as the $\exp(-1/g)$ dependence of the quark condensate on the QCD coupling $g$. Barrois then left academic physics. At around the same time the subject was also treated by David Bailin and Alexander Love at Sussex University, who studied various pairing patterns in detail, but did not give much attention to the confinement requirements and the phenomenology of color superconductivity in real-world quark matter.

Apart from papers by Masaharu Iwaskai and T. Iwado of Kochi University in 1995, there was little activity until 1998, when there was a major upsurge of interest in dense quark matter and color superconductivity, sparked by the simultaneously published work of two groups, one at the Institute for Advanced Study in Princeton and the other at Stony Brook University. These physicists pointed out that the strength of the strong interaction makes the phenomenon much more significant than had previously been suggested. These and other groups went on to investigate the combinatorial complexity of the many possible phases of color superconducting quark matter, and perform accurate calculations in the well-controlled limit of infinite density. Since then, interest in the topic has steadily grown, with current research (as of 2007) focusing on the detailed mapping of a plausible phase diagram for dense quark matter, and the search for observable signatures of the occurrence of these forms of matter in compact stars.

# Proximity Effect (Superconductivity)

Proximity effect or Holm-Meissner effect is a term used in the field of superconductivity to describe phenomena that occur when a superconductor (S) is placed in contact with a "normal" (N) non-superconductor. Typically the critical temperature of the superconductor is suppressed and signs of weak superconductivity are observed in the normal material over mesoscopic distances. The proximity effect is known since the pioneering work by R. Holm and W. Meissner. They have observed zero resistance in SNS pressed contacts, in which two superconducting metals are separated by a thin film of a non-superconducting (i.e. normal) metal. The discovery of the supercurrent in SNS contacts is sometimes mistakenly attributed to B. Josephson's 1962 work, yet the effect was known long before his publication and was understood as the proximity effect.

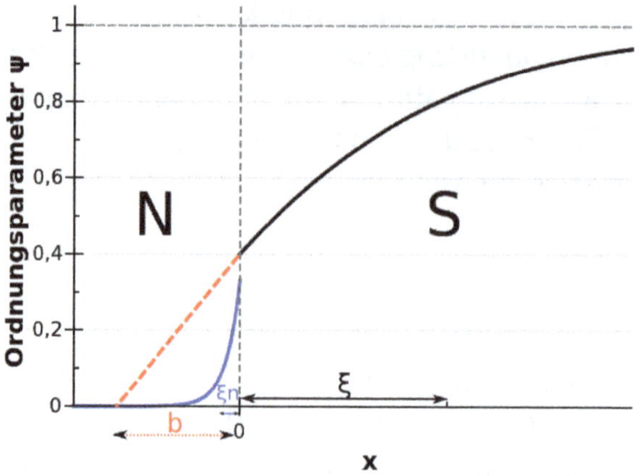


Plot showing superconducting electron density versus depth in normal and superconducting layers with two coherence lengths, $\xi$ and $\xi_n$.

## Origin of the Effect

Electrons in the superconducting state of a superconductor are ordered in a very different way than in a normal metal, i.e. they are paired into Cooper pairs. Furthermore, electrons in a material cannot be said to have a definitive position because of the momentum-position complementarity. In solid state physics one generally chooses a momentum space basis, and all electron states are filled with electrons until the Fermi surface in a metal, or until the gap edge energy in the superconductor.

Because of the nonlocality of the electrons in metals, the properties of those electrons cannot change infinitely quickly. In a superconductor, the electrons are ordered as superconducting Cooper pairs; in a normal metal, the electron order is gapless (single-electron states are filled up to the Fermi surface). If the superconductor and normal metal are brought together, the electron order in the one system cannot infinitely abruptly change into the other order at the border. Instead, the paired state in the superconducting layer is carried over to the normal metal, where the pairing is destroyed by scattering events, causing the Cooper pairs to lose their coherence. For very clean metals, such as copper, the pairing can persist for hundreds of microns.

Conversely, the (gapless) electron order present in the normal metal is also carried over to the superconductor in that the superconducting gap is lowered near the interface.

The microscopic model describing this behavior in terms of single electron processes is called Andreev reflection. It describes how electrons in one material take on the order of the neighboring layer by taking into account interface transparency and the states (in the other material) from which the electrons can scatter.

## Overview

As a contact effect, the proximity effect is closely related to thermoelectric phenomena like the Peltier effect or the formation of pn junctions in semiconductors. The proximity effect enhancement of $T_c$ is largest when the normal material is a metal with a large diffusivity rather than an insulator (I). Proximity-effect suppression of $T_c$ in a superconductor is largest when the normal material is ferromagnetic, as the presence of the internal magnetic field weakens superconductivity (Cooper pairs breaking).

## Research

The study of S/N, S/I and S/S' (S' is lower superconductor) bilayers and multilayers has been a particularly active area of SPE research. The behavior of the compound structure in the direction parallel to the interface differs from that perpendicular to the interface. In type II superconductors exposed to a magnetic field parallel to the interface, vortex defects will preferentially nucleate in the N or I layers and a discontinuity in behavior is observed when an increasing field forces them into the S layers. In type I superconductors, flux will similarly first penetrate N layers. Similar qualitative changes in behavior do not occur when a magnetic field is applied perpendicular to the S/I or S/N interface. In S/N and S/I multilayers at low temperatures, the long penetration depths and coherence lengths of the Cooper pairs will allow the S layers to maintain a mutual, three-dimensional quantum state. As temperature is increased, communication between the S layers is destroyed resulting in a crossover to two-dimensional behavior. The anisotropic behavior of S/N, S/I and S/S' bilayers and multilayers has served as a basis for understanding the far more complex critical field phenomena observed in the highly anisotropic cuprate high-temperature superconductors.

Recently the Holm-Meissner proximity effect was observed in graphene by the Morpurgo research group. The experiments have been done on nanometer scale devices made of single graphene layers with superimposed superconducting electrodes made of 10 nm Ti and 70 nm Al films. Al is a superconductor, which is responsible for inducing superconductivity into graphene. The distance between the electrodes was in the range between 100 nm and 500 nm. The proximity effect is manifested by observations of a supercurrent, i.e. a current flowing through the graphene junction with zero voltage on the junction. By using the gate electrodes the researches have shown that the proximity effect occurs when the carriers in the graphene are electrons as well as when the carriers are holes. The critical current of the devices was above zero even at the Dirac point.

## References


- Dew-Hughes, D. (1978). "Solid-state (bronze process) $V_3Ga$ from a V-Al alloy core". Journal of Applied Physics. 49: 327. Bibcode:1978JAP....49..327D. doi:10.1063/1.324390
- Usoskin, A., & Freyhardt, H. C. (2011). "YBCO-Coated Conductors Manufactured by High-Rate Pulsed Laser Deposition". MRS Bulletin. 29 (8): 583. doi:10.1557/mrs2004.165
- Russo, R. E., Reade, R. P., McMillan, J. M., & Olsen, B. L. (1990). "Metal buffer layers and Y-Ba-Cu-O thin films on Pt and stainless steel using pulsed laser deposition". Journal of Applied Physics. 68 (3): 1354. Bibcode:-1990JAP....68.1354R. doi:10.1063/1.346681
- "Characteristics of Superconducting Magnets". Superconductivity Basics. American Magnetics Inc. 2008. Retrieved 2008-10-11
- Beales, Timothy P.; Jutson, Jo; Le Lay, Luc; Mölgg, Michelé (1997). "Comparison of the powder-in-tube processing properties of two $(Bi_{2-x}Pb_x)Sr_2Ca_2Cu_3O_{10+\delta}$ powders". Journal of Materials Chemistry. 7 (4): 653. doi:10.1039/a606896k
- Holm, R.; Meissner, W. (1932). "Messungen mit Hilfe von flüssigem Helium. XIII". Z. Phys. 74: 715. doi:10.1007/bf01340420

# 2

# Features and Models of Superconductivity

Diamagnetism, Meissner effect, critical current and field and isotope effect are some of the features of superconductivity. The expulsion of a magnetic field from a superconductor is known as the Meissner effect. The chapter strategically encompasses and incorporates the major components and key concepts of superconductivity, providing a complete understanding.

## Characteristic Effects of Superconductivity

In this chapter, we will detail the various characteristic effects observed for superconducting materials. These have been documented over the decades following measurements of the electrical, magnetic, thermal and other properties.

As mentioned before, the resistance of metallic materials decreases with a decrease in temperature and displays a $T^5-$ behaviour at low-temperatures.

While the amount of residual resistivity (i.e., $\rho(T \to 0)$) does depend on the purity of the material, the temperature variation of $\rho$ is smooth. In contrast, for superconductors and as found by Onnes for the first time for the case of mercury, the dc resistivity decreases sharply to zero at the critical temperature $T_c$. A schematic variation of $\rho(T)$ for a non-superconducting (metallic) and a superconducting sample is shown in Figure. Note that in a typical measurement of dc resistivity, a constant dc current is applied to the sample and the voltage developed across it is measured as a function of temperature. On the other hand, for measuring the ac resistivity (or conductivity), an ac current source has to be used.

## Drude Model

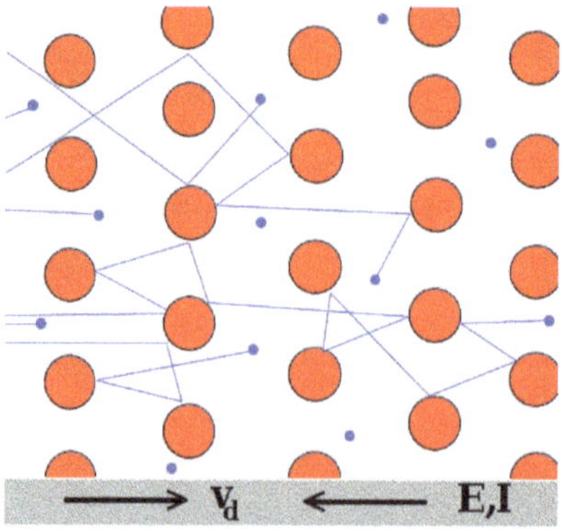


Drude model electrons (shown here in blue) constantly bounce between heavier, stationary crystal ions (shown in red).

The Drude model of electrical conduction was proposed in 1900 by Paul Drude to explain the transport properties of electrons in materials (especially metals). The model, which is an application of kinetic theory, assumes that the microscopic behavior of electrons in a solid may be treated classically and looks much like a pinball machine, with a sea of constantly jittering electrons bouncing and re-bouncing off heavier, relatively immobile positive ions.

The two most significant results of the Drude model are an electronic equation of motion,

$$\frac{d}{dt}\langle \boldsymbol{p}(t)\rangle = q\left(\boldsymbol{E} + \frac{\langle \boldsymbol{p}(t)\rangle \times \boldsymbol{B}}{m}\right) - \frac{\langle \boldsymbol{p}(t)\rangle}{\tau},$$

and a linear relationship between current density J and electric field E,

$$\boldsymbol{J} = \left(\frac{nq^2\tau}{m}\right)\boldsymbol{E}.$$

Here $t$ is the time, $\langle p\rangle$ is the average momentum per electron and $q$, $n$, $m$, and $\tau$ are respectively the electron charge, number density, mass, and mean free time between ionic collisions (that is, the mean time an electron has traveled since the last collision, not the average time between collisions). The latter expression is particularly important because it explains in semi-quantitative terms why Ohm's law, one of the most ubiquitous relationships in all of electromagnetism, should be true.

The model was extended in 1905 by Hendrik Antoon Lorentz (and hence is also known as the Drude–Lorentz model) and is a classical model. Later it was supplemented with the results of quantum theory in 1933 by Arnold Sommerfeld and Hans Bethe, leading to the Drude–Sommerfeld model.

## Assumptions

The Drude model considers the metal to be formed of a mass of positively charged ions from which a number of "free electrons" were detached. These may be thought to have become delocalized when the valence levels of the atom came in contact with the potential of the other atoms.

The Drude model neglects any long-range interaction between the electron and the ions or between the electrons. The only possible interaction of a free electron with its environment is via instantaneous collisions. The average time between subsequent collisions of such an electron is $\tau$, and the nature of the collision partner of the electron does not matter for the calculations and conclusions of the Drude model.

After a collision event, the velocity (and direction) of the electron only depends on the local temperature distribution and is completely independent of the velocity of the electron before the collision event.

## Explanations

### DC Field

The simplest analysis of the Drude model assumes that electric field E is both uniform and constant, and that the thermal velocity of electrons is sufficiently high such that they accumulate only an infinitesimal amount of momentum $dp$ between collisions, which occur on average every $\tau$ seconds.

Then an electron isolated at time $t$ will on average have been traveling for time $\tau$ since its last collision, and consequently will have accumulated momentum

$$\Delta\langle \boldsymbol{p}\rangle = q\boldsymbol{E}\tau.$$

During its last collision, this electron will have been just as likely to have bounced forward as backward, so all prior contributions to the electron's momentum may be ignored, resulting in the expression

$$\langle \boldsymbol{p}\rangle = q\boldsymbol{E}\tau.$$

Substituting the relations

$$\langle \boldsymbol{p}\rangle = m\langle \boldsymbol{v}\rangle,$$

$$\boldsymbol{J} = nq\langle \boldsymbol{v}\rangle,$$

results in the formulation of Ohm's law mentioned above:

$$\boldsymbol{J} = \left(\frac{nq^2\tau}{m}\right)\boldsymbol{E}.$$

## Time-varying Analysis

The dynamics may also be described by introducing an effective drag force. At time $t = t_0 + dt$ the average electron's momentum will be

$$\langle \boldsymbol{p}(t_0 + dt)\rangle = \left(1 - \frac{dt}{\tau}\right)\left(\langle \boldsymbol{p}(t_0)\rangle + q\boldsymbol{E}dt\right),$$

because, on average, a fraction of $1 - dt/\tau$ of the electrons will not have experienced another collision, and the ones that have will contribute to the total momentum to only a negligible order.

With a bit of algebra and dropping terms of order $dt^2$, this results in the differential equation

$$\frac{d}{dt}\langle \boldsymbol{p}(t)\rangle = q\boldsymbol{E} - \frac{\langle \boldsymbol{p}(t)\rangle}{\tau},$$

where $\langle p \rangle$ denotes average momentum and $q$ the charge of the electrons. This, which is an inhomogeneous differential equation, may be solved to obtain the general solution of

$$\langle \boldsymbol{p}(t)\rangle = q\tau\boldsymbol{E}(1 - e^{-t/\tau}) + \langle \boldsymbol{p}(0)\rangle e^{-t/\tau}$$

for $p(t)$. The steady state solution $d\langle P\rangle / dt = 0)$ is then

$$\langle \boldsymbol{p} \rangle = q\tau \boldsymbol{E}.$$

As above, average momentum may be related to average velocity and this in turn may be related to current density,

$$\langle \boldsymbol{p} \rangle = m\langle \boldsymbol{v} \rangle,$$

$$\boldsymbol{J} = nq\langle \boldsymbol{v} \rangle,$$

and the material can be shown to satisfy Ohm's law with a DC-conductivity $\sigma_0$:

$$\mathrm{J} = \left( \frac{nq^2\tau}{m} \right) \mathbf{E}.$$

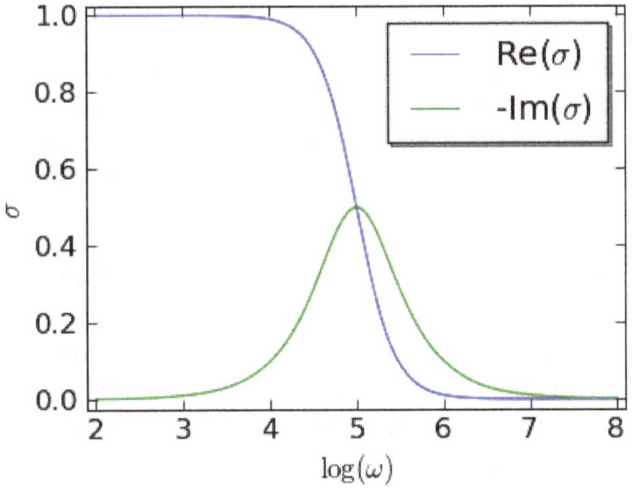


Complex conductivity for different frequencies assuming that $\tau = 10^{-5}$ and that $\sigma_0 = 1$.

The Drude model can also predict the current as a response to a time-dependent electric field with an angular frequency $\omega$, in which case

$$\sigma(\omega) = \frac{\sigma_0}{1 + i\omega\tau} = \frac{\sigma_0}{1 + \omega^2\tau^2} - i\omega\tau \frac{\sigma_0}{1 + \omega^2\tau^2}.$$

Here it is assumed that

$$E(t) = \Re(E_0 e^{i\omega t});$$

$$J(t) = \Re(\sigma(\omega) E_0 e^{i\omega t}).$$

In other conventions, $i$ is replaced by $-i$ in all equations. The imaginary part indicates that the current lags behind the electrical field, which happens because the electrons need roughly a time $\tau$ to accelerate in response to a change in the electrical field. Here the Drude model is applied to electrons; it can be applied both to electrons and holes; i.e., positive charge carriers in semiconductors. The curves for $\sigma(\omega)$ are shown in the graph.

## Drude Response in Real Materials

The characteristic behavior of a Drude metal in the time or frequency domain, i.e. exponential relaxation with time constant $\tau$ or the frequency dependence for $\sigma(\omega)$ stated above, is called Drude response. In a conventional, simple, real metal (e.g. sodium, silver, or gold at room temperature) such behavior is not found experimentally, because the characteristic frequency $\tau^{-1}$ is in the infrared frequency range, where other features that are not considered in the Drude model (such as band structure) play an important role. But for certain other materials with metallic properties, frequency-dependent conductivity was found that closely follows the simple Drude prediction for $\sigma(\omega)$. These are materials where the relaxation rate $\tau^{-1}$ is at much lower frequencies. This is the case for certain doped semiconductor single crystals, high-mobility two-dimensional electron gases, and heavy-fermion metals.

## Accuracy of the Model

Historically, the Drude formula was first derived in an incorrect way, namely by assuming that the charge carriers form an ideal gas. It is now known that they follow Fermi–Dirac distribution and have appreciable interactions, but amazingly, the result turns out to be the same as the Drude model because, as Lev Landau derived in 1957, a gas of interacting particles can be described by a system of almost non-interacting 'quasiparticles' that, in the case of electrons in a metal, can be well modelled by the Drude equation.

This simple classical Drude model provides a very good explanation of DC and AC conductivity in metals, the Hall effect, and thermal conductivity (due to electrons) in metals near room temperature. The model also explains the Wiedemann–Franz law of 1853. However, it greatly overestimates the electronic heat capacities of metals. In reality, metals and insulators have roughly the same heat capacity at room temperature. The model can be applied to positive (hole) charge carriers, as demonstrated by the Hall effect.

One note of trivia surrounding the theory is that in his original paper Drude made a conceptual error, estimating electrical conductivity to in fact be only half of what it classically should have been.

## Frequency Dependent Conductivity in the Drude Model

The Drude model treats the electron gas with the methods of the kinetic theory of a dilute neutral gas with the following assumptions: (i) between collisions, the motion of electrons is considered to be independent of the static ions or the other electrons, (ii) collisions are thought to be instantaneous events which abruptly change the velocities of the electrons, (iii) a relaxation time $\tau$ is introduced which is the mean time between collisions, and (iv) electrons achieve equilibrium with their surroundings only via collisions. The conductivity in a spatially uniform and time-independent electric field is given by $\sigma_0 = \frac{ne^2\tau}{m}$. For a time-dependent field represented by $\boldsymbol{E}(t) = \Re(\boldsymbol{E}(\omega)e^{-i\omega t})$, one obtains the current density $\boldsymbol{j}(\omega) = \sigma(\omega)\boldsymbol{E}(\omega)$ where the frequency dependent conductivity is given by

$$\sigma(\omega) = \frac{\sigma_0}{1 - i\omega\tau} = \frac{\sigma_0}{1 + \omega^2\tau^2} + i\frac{\sigma_0\omega\tau}{1 + \omega^2\tau^2}$$

## Two Fluid Model

Since the response of superconductors to ac fields is known to be dissipative, a simple two-fluid model is introduced to explain the ac conductivity of supercondcutors. Consider a material with total electron density $n$, which comprises of a superconducting part $n_s$ and a normal part $n_n$ having relaxation times $\tau_s$ and $\tau_n$ , respectively. Since infinite conductivity is obtained for a superconductor in the dc case, it is natural to consider $\tau_s = \infty$. The shape of the $\frac{\sigma_0}{1+\omega^2\tau^2}$ part of the conductivity as a function of frequency is bell shaped. At a given temperature, the height of the peak at zero frequency $(\sigma_0)$ grows as $\tau$ increases with the area under the curve remaining constant. Note that $\int \frac{d\omega}{1+\omega^2\tau^2} = \frac{\pi}{2\tau}$ and hence the area under the curve is $\frac{ne^2\pi}{2m}$. As $\tau \to \infty$, the width of the curve approaches zero and the height diverges with the area remaining constant (i.e., a Dirac-delta function). Similarly, the imaginary part of the conductivity approaches $\frac{ne^2}{m\omega}$ as $\tau \to \infty$.

Therefore, at non-zero frequencies, the superconducting fraction of the electrons contribute only to the imaginary part of the conductivity which is $\frac{n_s e^2}{m\omega}$. Further, in the limit $\omega\tau_n \ll 1$, the real part of the conductivity is approximately $\frac{n_n e^2 \tau}{m}$ which is finite. This illustrates the dissipative behaviour of a superconductor in an ac field.

Since a dc current can flow in a superconductor without dissipation, it is possible to set up currents in superconducting loops which do not decay with time. These are called persistent currents. Indeed, solenoid magnets made of superconducting wires are commercially available which can routinely provide magnetic fields as large as 100-200 kOe. The highest persistent field (268 kOe) has been achieved using cuprate based (YBCO) superconductors. In reality, there is a small decay of the circulating current due to something called flux flow resistance. This is however extremely small and only ppm level changes in the magnetic fields are seen over decades.

Two-fluid model is a macroscopic traffic flow model to represent traffic in a town/city or metropolitan area, put forward in the 1970s by Ilya Prigogine and Robert Herman.

There is also a two-fluid model which helps explain the behavior of superfluid helium. This model states that there will be two components in liquid helium below its lambda point (the temperature where superfluid forms). These components are a normal fluid and a superfluid component. Each liquid has a different density and together their sum makes the total density, which remains constant. The ratio of superfluid density to the total density increases as the temperature approaches absolute zero.

## Diamagnetism

Diamagnetic materials are repelled by a magnetic field; an applied magnetic field creates an induced magnetic field in them in the opposite direction, causing a repulsive force. In contrast, paramagnetic and ferromagnetic materials are attracted by a magnetic field. Diamagnetism is a quantum mechanical effect that occurs in all materials; when it is the only contribution to the magnetism the material is called diamagnetic. In paramagnetic and ferromagnetic substances the weak diamagnetic force is overcome by the attractive force of magnetic dipoles in the material. The magnetic permeability of diamagnetic materials is less than $\mu_0$, the permeability of vacuum. In most materials diamagnetism is a weak effect which can only be detected by sensitive laboratory instruments, but a superconductor acts as a strong diamagnet because it repels a magnetic field entirely from its interior.

Pyrolytic carbon has one of the largest diamagnetic constants of any room temperature material. Here a pyrolytic carbon sheet is levitated by its repulsion from the strong magnetic field of neodymium magnets.

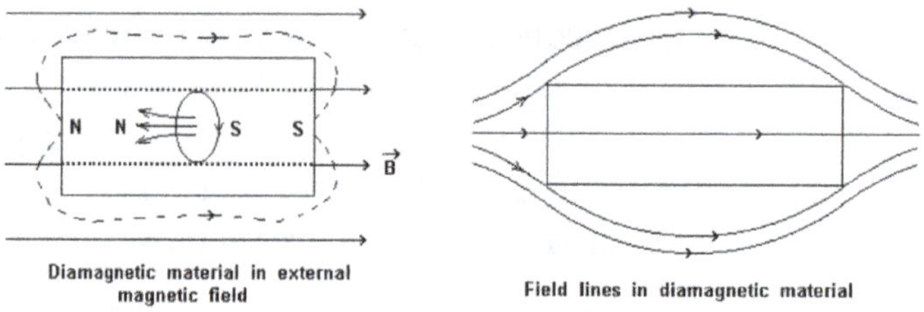


Dimagnetic material interaction in magnetic field.

Diamagnetism was first discovered when Sebald Justinus Brugmans observed in 1778 that bismuth and antimony were repelled by magnetic fields. In 1845, Michael Faraday demonstrated that it was a property of matter and concluded that every material responded (in either a diamagnetic or paramagnetic way) to an applied magnetic field. He adopted the term *diamagnetism* after it was suggested to him by William Whewell.

## Materials

| Notable diamagnetic materials | |
|---|---|
| **Material** | **$\chi_v$ (× $10^{-5}$)** |
| Superconductor | $-10^5$ |
| Pyrolytic carbon | −40.9 |
| Bismuth | −16.6 |

| Notable diamagnetic materials | |
|---|---|
| **Material** | **$\chi_v$ (× $10^{-5}$)** |
| Mercury | −2.9 |
| Silver | −2.6 |
| Carbon (diamond) | −2.1 |
| Lead | −1.8 |
| Carbon (graphite) | −1.6 |
| Copper | −1.0 |
| Water | −0.91 |

Diamagnetism, to a greater or lesser degree, is a property of all materials and always makes a weak contribution to the material's response to a magnetic field. For materials that show some other form of magnetism (such as ferromagnetism or paramagnetism), the diamagnetic contribution becomes negligible. Substances that mostly display diamagnetic behaviour are termed diamagnetic materials, or diamagnets. Materials called diamagnetic are those that laymen generally think of as *non-magnetic*, and include water, wood, most organic compounds such as petroleum and some plastics, and many metals including copper, particularly the heavy ones with many core electrons, such as mercury, gold and bismuth. The magnetic susceptibility values of various molecular fragments are called Pascal's constants.

Diamagnetic materials, like water, or water-based materials, have a relative magnetic permeability that is less than or equal to 1, and therefore a magnetic susceptibility less than or equal to 0, since susceptibility is defined as $\chi_v = \mu_v - 1$. This means that diamagnetic materials are repelled by magnetic fields. However, since diamagnetism is such a weak property, its effects are not observable in everyday life. For example, the magnetic susceptibility of diamagnets such as water is $\chi_v = -9.05\times10^{-6}$. The most strongly diamagnetic material is bismuth, $\chi_v = -1.66\times10^{-4}$, although pyrolytic carbon may have a susceptibility of $\chi_v = -4.00\times10^{-4}$ in one plane. Nevertheless, these values are orders of magnitude smaller than the magnetism exhibited by paramagnets and ferromagnets. Note that because $\chi_v$ is derived from the ratio of the internal magnetic field to the applied field, it is a dimensionless value.

All conductors exhibit an effective diamagnetism when they experience a changing magnetic field. The Lorentz force on electrons causes them to circulate around forming eddy currents. The eddy currents then produce an induced magnetic field opposite the applied field, resisting the conductor's motion.

## Superconductors

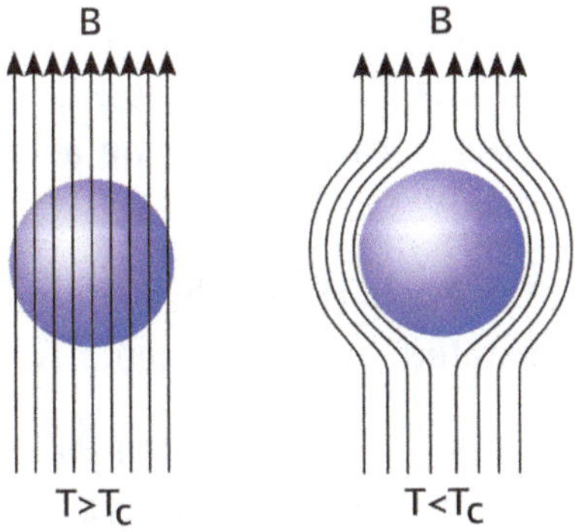

Transition from ordinary conductivity (left) to superconductivity (right). At the transition, the superconductor expels the magnetic field and then acts as a perfect diamagnet.

Superconductors may be considered perfect diamagnets ($\chi_v = -1$), because they expel all fields (except in a thin surface layer) due to the Meissner effect.

## Demonstrations

### Curving Water Surfaces

If a powerful magnet (such as a supermagnet) is covered with a layer of water (that is thin compared to the diameter of the magnet) then the field of the magnet significantly repels the water. This causes a slight dimple in the water's surface that may be seen by its reflection.

### Levitation

A live frog levitates inside a 32 mm (1.26 in) diameter vertical bore of a Bitter solenoid in a magnetic field of about 16 teslas at the Nijmegen High Field Magnet Laboratory.

Diamagnets may be levitated in stable equilibrium in a magnetic field, with no power consumption. Earnshaw's theorem seems to preclude the possibility of static magnetic levitation. However, Earnshaw's theorem applies only to objects with positive susceptibilities, such as ferromagnets (which have a permanent positive moment) and paramagnets (which induce a positive moment). These are attracted to field maxima, which do not exist in free space. Diamagnets (which induce a negative moment) are attracted to field minima, and there can be a field minimum in free space.

A thin slice of pyrolytic graphite, which is an unusually strong diamagnetic material, can be stably floated in a magnetic field, such as that from rare earth permanent magnets. This can be done with all components at room temperature, making a visually effective demonstration of diamagnetism.

The Radboud University Nijmegen, the Netherlands, has conducted experiments where water and other substances were successfully levitated. Most spectacularly, a live frog was levitated.

In September 2009, NASA's Jet Propulsion Laboratory in Pasadena, California announced it had successfully levitated mice using a superconducting magnet, an important step forward since mice are closer biologically to humans than frogs. JPL said it hopes to perform experiments regarding the effects of microgravity on bone and muscle mass.

Recent experiments studying the growth of protein crystals have led to a technique using powerful magnets to allow growth in ways that counteract Earth's gravity.

A simple homemade device for demonstration can be constructed out of bismuth plates and a few permanent magnets that levitate a permanent magnet.

## Theory

The electrons in a material generally circulate in orbitals, with effectively zero resistance and act like current loops. Thus it might be imagined that diamagnetism effects in general would be very, very common, since any applied magnetic field would generate currents in these loops that would oppose the change, in a similar way to superconductors, which are essentially perfect diamagnets. However, since the electrons are rigidly held in orbitals by the charge of the protons and are further constrained by the Pauli exclusion principle, many materials exhibit diamagnetism, but typically respond very little to the applied field.

The Bohr–van Leeuwen theorem proves that there cannot be any diamagnetism or paramagnetism in a purely classical system. However, the classical theory for Langevin diamagnetism gives the same prediction as the quantum theory. The classical theory is given below.

## Langevin Diamagnetism

The Langevin theory of diamagnetism applies to materials containing atoms with closed shells. A field with intensity $B$, applied to an electron with charge $e$ and mass $m$, gives rise to Larmor precession with frequency $\omega = eB/2m$. The number of revolutions per unit time is $\omega/2\pi$, so the current for an atom with $Z$ electrons is (in SI units)

$$I = -\frac{Ze^2B}{4\pi m}.$$

The magnetic moment of a current loop is equal to the current times the area of the loop. Suppose the field is aligned with the $z$ axis. The average loop area can be given as $\pi\left\langle \rho^2 \right\rangle$, where $\left\langle \rho^2 \right\rangle$ is the mean square distance of the electrons perpendicular to the $z$ axis. The magnetic moment is therefore

$$\mu = -\frac{Ze^2B}{4m}\left\langle \rho^2 \right\rangle.$$

If the distribution of charge is spherically symmetric, we can suppose that the distribution of $x,y,z$ coordinates are independent and identically distributed. Then $\left\langle x^2 \right\rangle = \left\langle y^2 \right\rangle = \left\langle z^2 \right\rangle = \frac{1}{3}\left\langle r^2 \right\rangle$, where $\left\langle r^2 \right\rangle$ is the mean square distance of the electrons from the nucleus. Therefore, $\left\langle \rho^2 \right\rangle = \left\langle x^2 \right\rangle + \left\langle y^2 \right\rangle = \frac{2}{3}\left\langle r^2 \right\rangle$. If $N$ is the number of atoms per unit volume, the diamagnetic susceptibility in SI units is

$$\chi = \frac{\mu_0 N\mu}{B} = -\frac{\mu_0 NZe^2}{6m}\left\langle r^2 \right\rangle.$$

## In Metals

The Langevin theory does not apply to metals because they have non-localized electrons. The theory for the diamagnetism of a free electron gas is called Landau diamagnetism and instead considers the weak counter-acting field that forms when their trajectories are curved due to the Lorentz force. Landau diamagnetism, however, should be contrasted with Pauli paramagnetism, an effect associated with the polarization of delocalized electrons' spins. For the bulk case of a 3D system and low magnetic fields it can be calculated using Landau quantization:

$$\chi = -\frac{e^2}{12\pi^2 m\hbar}\sqrt{2mE_F}.$$

where $E_F$ is the Fermi energy. This is equivalent to $-1/3\mu_B^2 g(E_F)$ exactly minus a third of Pauli paramagnetic susceptibility, where $\mu_B = e\hbar/(2m)$ is the Bohr magneton and $g(E)$ in the density of states.

We note here that the response of a superconductor to an applied magnetic field is distinctly different from what one might expect of a hypothetical perfect or ideal conductor ($\sigma = \infty$). Imagine that a normal, non-superconducting material were cooled in the absence of a magnetic field. Further consider that this material becomes a perfect conductor below some temperature. If a magnetic field is now turned on, the magnetic field $(B)$ inside the conductor should remain unchanged at zero (this follows from Maxwell equations and is shown below). On the other hand, if the material were cooled in the presence of a magnetic field, the field would penetrate the material in the finite conductivity state and will continue to do so when the material is cooled to temperatures where it has $\sigma = \infty$. In contrast to the above behaviour expected of a perfect or ideal conductor, a superconductor (for $T < T_c$) always has $B = 0$ inside it, i.e., whether it is cooled in a magnetic field or the field is turned on after cooling the sample (issues such as penetration depth and critical fields will be discussed later). The expulsion of field from the superconductor when it is cooled in a field is called the Meissner effect while a similar consequence which takes place when the sample is cooled in zero field and then a field is applied for $T < T_c$ is called perfect diamagnetism. Let us start from the Maxwell equation also referred to as the Faraday law

$$\vec{\nabla} \times \vec{E} = -\frac{1}{c}\frac{\partial \vec{B}}{\partial t}$$

Newton's law for electrons under the influence of an electric field is given by (note that this is strictly valid only for spatially uniform fields which is not the case at interfaces)

$$m\frac{d\vec{v}}{dt} = -e\vec{E}$$

$$\Rightarrow m\frac{\partial(-ne\vec{v})}{\partial t} = ne^2\vec{E}$$

$$\Rightarrow \vec{\nabla} \times \vec{E} = \frac{m}{ne^2}\vec{\nabla} \times \frac{\partial \vec{J}}{\partial t}$$

Consequently

$$\vec{\nabla} \times \frac{\partial \vec{J}}{\partial t} = -\frac{ne^2}{mc} \frac{\partial \vec{B}}{\partial t}$$

Making use of another of Maxwell equations (neglecting the displacement current and taking $B = H$ for the superconductor we get $\nabla \times B = \frac{4\pi}{c} J$), we obtain

$$\vec{\nabla} \times \frac{\partial \left( \vec{\nabla} \times \vec{B} \right)}{\partial t} = -\frac{4\pi ne^2}{mc^2} \frac{\partial \vec{B}}{\partial t}$$

Further, using the vector identity $\vec{\nabla} \times \vec{\nabla} \times \vec{B} = \vec{\nabla}(\vec{\nabla} \cdot \vec{B}) - \nabla^2 \vec{B}$ along with no-magnetic-monopole condition $\nabla \cdot B = 0$,

$$\nabla^2 \frac{\partial \vec{B}}{\partial t} = \frac{4\pi ne^2}{mc^2} \frac{\partial \vec{B}}{\partial t}$$

Applying the above equation to a slab of a perfect conductor implies that $\frac{\partial B}{\partial t}$ must fall off exponentially inside the conductor with a characteristic length scale (London penetration depth $\lambda$) of $\sqrt{\frac{mc^2}{4\pi ne^2}}$. Therefore, changes in the field are attenuated exponentially inside a perfect conductor.

## London Moment

The London moment (after Fritz London) is a quantum-mechanical phenomenon whereby a spinning superconductor generates a magnetic field whose axis lines up exactly with the spin axis. The term may also refer to the magnetic moment of any rotation of any superconductor, caused by the electrons lagging behind the rotation of the object, although the field strength is independent of the charge carrier density in the superconductor.

### Gravity Probe B

A magnetometer determines the orientation of the generated field, which is interpolated to determine the axis of rotation. Gyroscopes of this type can be extremely accurate and stable. For example, those used in the Gravity Probe B experiment measured changes in gyroscope spin axis orientation to better than 0.5 milliarcseconds ($1.4 \times 10^{-7}$ degrees) over a one-year period. This is equivalent to an angular separation the width of a human hair viewed from 32 kilometers (20 miles) away.

The GP-B gyro consists of a nearly-perfect spherical rotating mass made of fused quartz, which provides a dielectric support for a thin layer of niobium superconducting material. To eliminate friction found in conventional bearings, the rotor assembly is centered by the electric field from six electrodes. After the initial spin-up by a jet of helium which brings the rotor to 4,000 RPM, the polished gyroscope housing is evacuated to an ultra-high vacuum to further reduce drag on the rotor. Provided the suspension electronics remain powered, the extreme rotational symmetry, lack of friction, and low drag will allow the angular momentum of the rotor to keep it spinning for about 15,000 years.

A sensitive DC SQUID magnetometer able to discriminate changes as small as one quantum, or about $2 \times 10^{-15}$ Wb, is used to monitor the gyroscope. A precession, or tilt, in the orientation of the rotor causes the London moment magnetic field to shift relative to the housing. The moving field passes through a superconducting pickup loop fixed to the housing, inducing a small electric current. The current produces a voltage across a shunt resistance, which is resolved to spherical coordinates by a microprocessor. The system is designed to minimize Lorentz torque on the rotor.

## Magnetic Field Strength

The magnetic field strength associated with a rotating superconductor is given by:

$$B = -\frac{2M}{Q}\omega$$

where $M$ and $Q$ are the mass and the charge of the superconducting charge carriers respectively. For the case of Cooper pairs of electrons, $M=2m_e$ and $Q=2e$. Despite the electrons existing in a strongly-interacting environment, $m_e$ denotes here the mass of the bare electrons (as in vacuum), and not e.g. the effective mass of conducting electrons of the normal phase.

## London Equations

In order to explain the Meissner effect in superconductors, Fritz and Heinz London proposed the following equations in 1935.

$$\vec{\nabla} \times \vec{J} = -\frac{ne^2}{mc}\vec{B} \Rightarrow \vec{J} = -\frac{ne^2\vec{A}}{mc}$$

$$\frac{\partial \vec{J}}{\partial t} = -\frac{ne^2}{m}\vec{E}$$

The second of the above equations is, in any case, valid for a perfect conductor while the first one helps explain Meissner effect. Note that the first equation did not follow from any micrscopic theory but was merely an attempt to describe empirical observations. The penetration depth varies with temperature in a qualitative manner as indicated in the figure below. The empirical variation has been found to follow

$$\lambda(T) = \frac{\lambda(0)}{\sqrt{1-(\frac{T}{T_c})^4}}$$

The values of $\lambda(0)$ range from tens of $\AA$ to thousands of $\AA$.

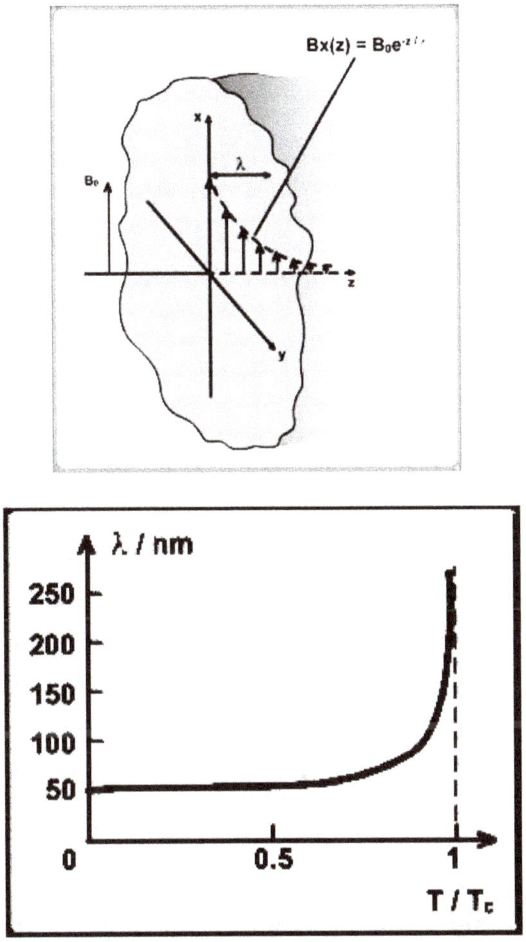


The figure illustrates the penetration of magnetic field inside a superconductor. The schematic variation of penetration depth with temperature is also shown.

The London equations, developed by brothers Fritz and Heinz London in 1935, relate current to electromagnetic fields in and around a superconductor. Arguably the simplest meaningful description of superconducting phenomena, they form the genesis of almost any modern introductory text on the subject. A major triumph of the equations is their ability to explain the Meissner effect, wherein a material exponentially expels all internal magnetic fields as it crosses the superconducting threshold.

## Formulations

There are two London equations when expressed in terms of measurable fields:

$$\frac{\partial \mathbf{j}_s}{\partial t} = \frac{n_s e^2}{m}\mathbf{E}, \qquad \nabla \times \mathbf{j}_s = -\frac{n_s e^2}{m}\mathbf{B}.$$

Here $\mathbf{j}_s$ is the superconducting current density, E and B are respectively the electric and magnetic fields within the superconductor, e is the charge of an electron & proton, m is electron mass, and $n_s$ is a phenomenological constant loosely associated with a number density of superconducting carriers. Throughout this chapter SI units are employed.

On the other hand, if one is willing to abstract away slightly, both the expressions above can more neatly be written in terms of a single "London Equation" in terms of the vector potential A:

$$\mathbf{j}_s = -\frac{n_s e^2}{m}\mathbf{A}.$$

The last equation suffers from only the disadvantage that it is not gauge invariant, but is true only in the Coulomb gauge, where the divergence of A is zero. This equation holds for magnetic fields that vary slowly in space.

## London Penetration Depth

If the second of London's equations is manipulated by applying Ampere's law,

$$\nabla \times \mathbf{B} = \mu_0 \mathbf{j},$$

then the result is the differential equation

$$\nabla^2 \mathbf{B} = \frac{1}{\lambda^2}\mathbf{B}, \qquad \lambda \equiv \sqrt{\frac{m}{\mu_0 n_s e^2}}.$$

Thus, the London equations imply a characteristic length scale, λ, over which external magnetic fields are exponentially suppressed. This value is the London penetration depth.

For an example, consider a superconductor within free space where the magnetic field outside the superconductor is a constant value pointed parallel to the superconducting boundary plane in the *z* direction. If *x* leads perpendicular to the boundary then the solution inside the superconductor may be shown to be

$$B_z(x) = B_0 e^{-x/\lambda}.$$

From here the physical meaning of the London penetration depth can perhaps most easily be discerned.

## Rationale for the London Equations

### Original Arguments

While it is important to note that the above equations cannot be formally derived, the Londons did follow a certain intuitive logic in the formulation of their theory. Substances across a stunningly wide range of composition behave roughly according to Ohm's law, which states that current is proportional to electric field. However, such a linear relationship is impossible in a superconductor for, almost by definition, the electrons in a superconductor flow with no resistance whatsoever. To this end, the London brothers imagined electrons as if they were free electrons under the influence of a uniform external electric field. According to the Lorentz force law

$$\mathbf{F} = e\mathbf{E} + e\mathbf{v} \times \mathbf{B}$$

these electrons should encounter a uniform force, and thus they should in fact accelerate uniformly. This is precisely what the first London equation states.

To obtain the second equation, take the curl of the first London equation and apply Faraday's law,

$$\nabla \times \mathrm{E} = -\frac{\partial \mathrm{B}}{\partial t},$$

to obtain

$$\frac{\partial}{\partial t}\left(\nabla \times \mathbf{j}_s + \frac{n_s e^2}{m}\mathbf{B}\right) = 0.$$

As it currently stands, this equation permits both constant and exponentially decaying solutions. The Londons recognized from the Meissner effect that constant nonzero solutions were nonphysical, and thus postulated that not only was the time derivative of the above expression equal to zero, but also that the expression in the parentheses must be identically zero. This results in the second London equation.

### Canonical Momentum Arguments

It is also possible to justify the London equations by other means. Current density is defined according to the equation

$$\mathbf{j}_s = -n_s e \mathbf{v}.$$

Taking this expression from a classical description to a quantum mechanical one, we must replace values j and v by the expectation values of their operators. The velocity operator

$$\mathbf{v} = \frac{1}{m}(\mathbf{p} - e\mathbf{A})$$

is defined by dividing the gauge-invariant, kinematic momentum operator by the particle mass $m$. We may then make this replacement in the equation above. However, an important assumption from the microscopic theory of superconductivity is that the superconducting state of a system is the ground state, and according to a theorem of Bloch's, in such a state the canonical momentum p is zero. This leaves

$$\mathbf{j}_s = -\frac{n_s e_s^2}{m}\mathbf{A},$$

which is the London equation according to the second formulation above.

## Critical Field

Superconductivity is characterized both by perfect conductivity (zero resistance) and by the expulsion of magnetic fields (the Meissner effect). Changes in either temperature or magnetic field can cause the phase transition between normal and superconducting states. For a given temperature, the highest magnetic field under which a material remains superconducting is known as the critical field. The highest temperature under which the superconducting state is seen is known as the critical temperature. At that temperature even the smallest external magnetic field will destroy the superconducting state, so the critical field is zero. As temperature decreases, the critical field increases generally to a maximum at absolute zero.

For a Type I superconductor the discontinuity in heat capacity seen at the superconducting transition is generally related to the slope of the critical field ($H_c$) at the critical temperature ($T_c$):

$$C_{super} - C_{normal} = \frac{T}{4\pi}\left(\frac{dH_c}{dT}\right)^2_{T=T_c}$$

There is also a direct relation between the critical field and the critical current - the maximum electric current density that a given superconducting material can carry, before switching into the normal state. According to Ampère's law any electric current induces a magnetic field, but superconductors exclude that field. On a microscopic scale the magnetic field is not quite zero at the edges of any given sample - a penetration depth applies. For a type I superconductor, the current must remain zero within the superconducting material (to be compatible with zero magnetic field), but can then go to non-zero values at the edges of the material on this penetration-depth length-scale, as the magnetic field rises. As long as the induced magnetic field at the edges is less than the critical field, the material remains superconducting, but at higher currents the field becomes too strong and the superconducting state is lost. This limit on current density has important practical implications in applications of superconducting materials - despite zero resistance they cannot carry unlimited quantities of electric power.

The geometry of the superconducting sample complicates practical measurement of the critical field - the critical field is defined for a cylindrical sample with the field parallel to the axis of symmetry. Under other conditions, for example for a spherical sample, there may be a mixed state with partial penetration of the magnetic field (and thus partial normal state) while a portion of the sample remains superconducting.

Type II superconductors allow a different sort of mixed state, where the magnetic field (above the lower critical field $H_{c1}$) is allowed to penetrate along cylindrical "holes" through the material, each of which carries a magnetic flux quantum. Along these flux cylinders the material is essentially in a normal, non-superconducting state, surrounded by superconductor where the magnetic field goes back to zero. The width of each cylinder is on the order of the penetration depth for the material. As the magnetic field increases, the flux cylinders move closer together and eventually at the upper critical field $H_{c2}$, they leave no room for the superconducting state and the zero-conductivity property is lost.

## Upper Critical Field

The upper critical field (UCF) is the magnetic field (usually expressed in teslas (T)) which completely suppresses superconductivity in a *Type II* superconductor at oK (absolute zero).

More properly, the UCF is a function of temperature (and pressure) and if these are not specified absolute zero and standard pressure are implied.

Werthamer–Helfand–Hohenberg theory predicts the upper critical field ($H_{c2}$) at 0 K from $T_c$ and the slope of $H_{c2}$ at $T_c$.

The UCF (at 0 K) can also be estimated from the coherence length(ξ) using the Ginzburg-Landau expression : $H_{c2} = 2.07 \times 10^{-15}\ Tm^2/(2\ Pi\ \xi^2)$.

Reference materials on superconductivity use $H_{c2}$ or $B_{c2}$ interchangeably since the materials are often exhibit perfect diamagnetism (with susceptibility, $\chi = -1$, hence giving equal magnitudes for $|H_{c2}|$ and $|B_{c2}|$).

## Lower Critical Field

The magnetic field at which the magnetic flux starts to penetrate a type-2 superconductor.

This topic discusses the effect of a magnetic field on the $T_c$ of a superconductor. The $T_c$ is found to decrease when a magnetic field is present and superconductivity can be completely suppressed by applying a sufficiently strong magnetic field (called the critical field) $H_c$. As will be discussed later, Type I superconductors have a single critical field $(H_c)$ while Type II superconductors have a lower and an upper critical field ($H_{c1}$ and $H_{c2}$, respectively) between which the substance is said to be in the mixed state. Typical values of $H_c$ are in the region of 100 Oe while those of $H_{c2}$ can be as high as 100 Tesla. Clearly, the Type II superconductors with high $H_{c2}$ are useful from the point of view of applications. The typical variation of $H_c$ with temperature is shown in the Figure below. It follows

$$H_c(T) = H_c(0)(1-(\frac{T}{T_c})^2)$$

Here, at a temperature $T$ a field of $H_c(T)$ is needed to suppress superconductivity completely.

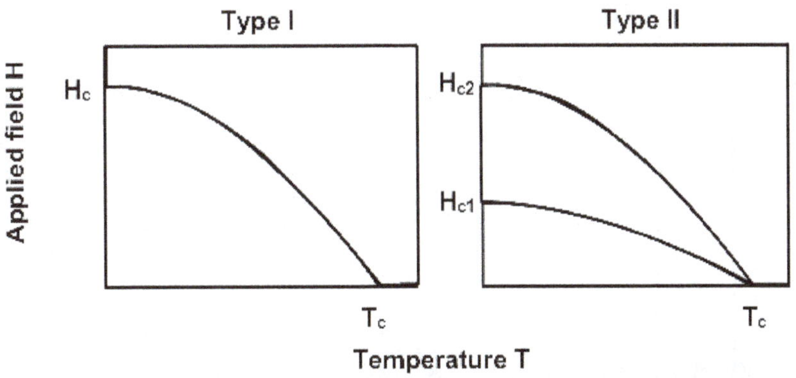


The schematic variation of critical field with temperature for Type-I and Type-II superconductors is shown in the figure. When the mouse is moved over different regions, it will tell you the state of the material.

## Critical Current

Likewise, there is a limit to the current that one can pass through a superconductor beyond which it is normal (non-superconducting). For a cylindrical conductor, one can estimate it in the following manner. For a current $I$ flowing uniformly through a cylindrical conductor of radius $a$, the field at the surface is $H = \frac{I}{2\pi a}$. Clearly, if this exceeds $H_c$ the whole wire will become normal. The critical current is therefore $I_c = 2\pi a H_c$. In practice, a current of hundreds of Amperes can be passed through superconducting wires without dissipation.

## Isotope Effect

Among the parameters which affect $T_c$, the isotopic mass $M$ plays an important role and changes $T_c$ as $T_c \propto M^{-\alpha}$ with $\alpha \approx 0.5$. The isotope effect was discovered in 1950 in Hg and Sn where different isotopes could be synthesized using methods of nuclear physics. This must have been an important clue in the development of the BCS theory since it indicated the importance of the lattice for the mechanism of superconductivity. The most important consequence of London equations is the screening of a magnetic field from the bulk of a superconductor. Consider the following geometries for which we will determine the field variation inside the superconductor.

## Solution of London Equations for Sample Cases

### Flat Slab in a Magnetic Field

Consider a flat superconducting slab of thickness $d$ in a magnetic field $H_a$ parallel to the slab. The boundary condition is that field match at $x = \pm d/2$. Subject to this condition, the solution to the London equation $\nabla^2 h = \frac{1}{\lambda^2} h$, where $h$ is the microscopic value of the flux density, is easily determined to be a superposition of two exponentials. The result can be written as

$$h = H_a \frac{\cosh(x/\lambda)}{\cosh(d/2\lambda)}$$

Clearly, the minimum value of the flux density is attained at the mid-plane of the slab where it has a value $H_a / \cosh(d/2\lambda)$.

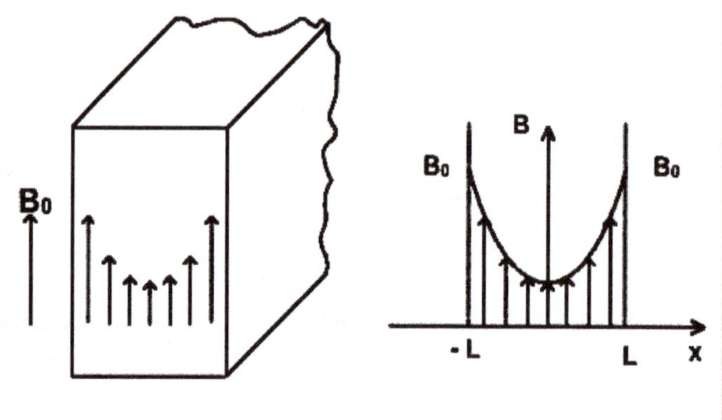


Cross section of a superconducting slab in a magnetic field

The field variation inside a superconducting slab is shown.

Averaging this internal field over the sample thickness one gets

$$B \equiv \bar{h} = H_a \frac{2\lambda}{d} \tanh\left(\frac{d}{2\lambda}\right)$$

Let us consider the limit $d \gg \lambda$. This leads to $B = 0$ deep inside the superconductor. In the other limit, i.e., $d \ll \lambda$, we expand $\tanh(x) = x - x^3/3$. Therefore, $B$ approaches $H_a(1 - \frac{d^2}{12\lambda^2})$. Since $B = H_a + 4\pi M$, we get

$$M = -\frac{H_a}{4\pi}\left(\frac{d^2}{12\lambda^2}\right)$$

As a consequence, magnetisation measurements can be made on thin films of known thicknesses and the penetration depth can be estimated from such measurements. Since the magnetisation is reduced below its Meissner value, the effective critical field for a thin sample is greater than that for bulk. The difference in the free energy between the normal state and the superconducting state is

$$(F_n - F_s)|_{H=0} = -\int_0^{H_c} M dH$$

For the case of complete flux expulsion, the above difference in free energies is

$$\Delta F = -\int_0^{H_c} \frac{H}{4\pi} dH = \frac{H_c^2}{8\pi}$$

This energy, which stabilises the superconducting state is called the condensation energy and $H_c$ is called the thermodynamic critical field.

For a thin film sample (with a field applied parallel to the plane) we get,

$$\Delta F = \frac{1}{4\pi}\frac{d^2}{12\lambda^2}\int_0^{H_{c\,par}} HdH \frac{H_{c\,par}^2}{8\pi}\frac{d^2}{12\lambda^2}$$

In terms of the bulk thermodynamic critical field

$$\frac{H_{c\,par}^2}{8\pi}\frac{d^2}{12\lambda^2} = \frac{H_c^2}{8\pi} or H_{c\,par} = \frac{\sqrt{12}\lambda}{d} H_c$$

## Critical Current of a Wire

Consider a long superconducting wire having a circular cross-section of radius $a$. Also, assume that $\lambda \gg a$. A current $I$ is passed through the wire. This gives rise to a circumferential magnetic field at the surface of the wire $H = \frac{2I}{ca}$. In a simple minded picture, when this field reaches $H_c$, the wire will become normal. Therefore, the critical current $I_c = \frac{ca}{2} H_c$ depends linearly on the radius and not on the area. The current flows only in a surface layer of thickness $\lambda$. Hence, the current density $J_c \approx \frac{I_c}{2\pi a\lambda} = \frac{H_c}{2\pi a\lambda}\frac{ca}{2}$. Therefore, $J_c = \frac{c}{4\pi}\frac{H_c}{\lambda}$.

## Free Energy Calculations

Now consider the case of a type I superconductor in a relatively large field. First we will carry out some calculations assuming zero demagnetisation factor. For a normal sample of volume $V$ in a magnetic field $H_a$, the Helmholtz free energy is given by

$$F_n = Vf_{n0} = V\frac{H_a^2}{8\pi} + V_{ext}\frac{H_a^2}{8\pi}$$

Here $f_{n0}$ is the free energy density in zero applied field. $V_{ext}$ is the volume external to the sample volume where the field $H_a$ exists. On the other hand, for a superconductor, the field is excluded from its interior and hence its free energy is given by

$$F_s = Vf_{s0} + V_{ext}\frac{H_a^2}{8\pi}$$

Here we have ignored the fact that the field actually penetrates in a layer of depth $\sim\lambda$ from the surface. The difference between the above two free energies is then

$$F_n - F_s = V(f_{n0} - f_{s0}) + V\frac{H_a^2}{8\pi}$$

Since the condensation energy density is the stabilisation energy

$$F_n - F_s = V\frac{H_c^2}{8\pi} + V\frac{H_a^2}{8\pi}$$

For $H_a = H_c$

$$(F_n - F_s)|_{H_c} = V\frac{H_c^2}{4\pi}$$

This is the energy increase (sample plus the surroundings) when a sample becomes normal at $H_a = H_c$. The increase comes about because the energy source (generator) maintaining the constant field does work against the back emf. This emf is induced as the flux threading the sample changes (starts entering the bulk of the superconductor). Actually, discussion in terms of a Helmholtz free energy is appropriate for a situation where $B$ is held constant (i.e., no induced emf). Here, we are holding $H$ constant so the appropriate thermodynamic potential is the Gibbs free energy $G$. Recall that the Gibbs free energy density $g$ is related to the Helmholtz free energy density $f$ as follows

$$g = f - \frac{hH}{4\pi}$$

In the normal state, the local flux density $h$ is equal to the average flux density $B$ which is the same as the applied field $H_a$. Therefore we get

$$G_n = Vf_{n0} - V\frac{H_a^2}{8\pi} - V_{ext}\frac{H_a^2}{8\pi}$$

In the superconducting state, flux is excluded from the superconductor, so $h = B = 0$. This gives

$$G_s = Vf_{s0} - V_{ext}\frac{H_a^2}{8\pi}$$

The difference between the two free energies is then

$$G_n - G_s = V(f_{n0} - f_{s0}) - V\frac{H_a^2}{8\pi}$$

For an applied field equal to the thermodynamic critical field, we get $G_n = G_s$ i.e., there will be a phase equilibrium between the normal and the superconducting phase at $H_a = H_c$.

## Field Variation for a Non-zero Demagnetisation Factor

Consider a spherical sample of radius $R$. Outside the sphere, we have $\nabla \bullet B = 0$ and $\nabla \times B = 0$. Consequently $\nabla^2 B = 0$. Clearly, $B$ approaches $H_a$ as $r \to \infty$. Also, the perpendicular component of $B$ is zero at $r = R$.The solution to to $\nabla^2 B = 0$ is then

$$B_{out} = H_a + \frac{H_a R^3}{2}\vec{\nabla}\left(\frac{\cos\theta}{r^2}\right)$$

Therefore, the tangential component of $B$ at the surface of the sphere can be calculated and is given below

$$B_\theta(r = R) = \frac{3}{2}H_a \sin\theta$$

This exceeds the applied field at the equator. Even when the applied field is less than $H_c$, so long as it is greater than $\frac{2}{3}H_c$, $B$ can attain a value of $H_c$ at the equator. Therefore, for $\frac{2}{3}H_c < H_a < H_c$ here will be a coexistence of normal and superconducting regions. This has been called the "intermediate" state. Note that this is different from the "mixed" state which occurs at applied magnetic fields between $H_{c1}$ and $H_{c2}$, even in the absence of demagnetisation effects.In general, for ellipsoidal samples (i.e., where a demagnetisation factor is well defined), when the applied field is in the range $1 - \eta < \frac{H_a}{H_c} < 1$ (where $\eta$ is the demagnetisation factor) an intermediate state will occur. The value of $\eta$ for a sphere is $1/3$, for a flat plate with field perpendicular to it is 1, for a long cylinder with the field along the axis it is $0$, and for a long cylinder with a field perpendicular to its axis it is $1/2$.

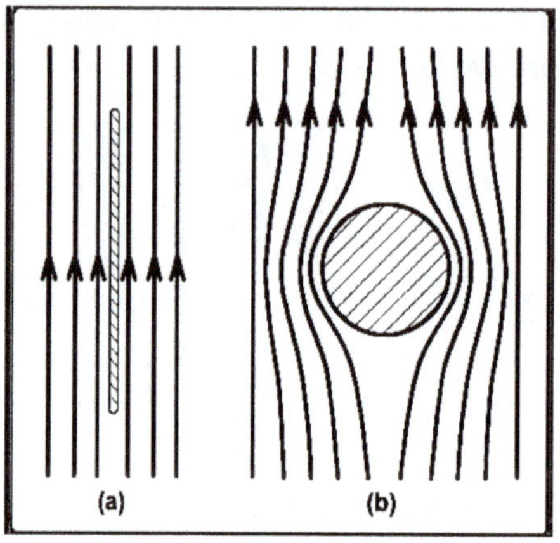


Contrast of exterior-field pattern (a) when demagnetizing coefficient is nearly zero and (b) when it is 1/3 for a sphere. In (b) the equatorial field is three halves the applied field for the case of full Meissner effect, which is shown.

The field pattern inside a spherically shaped superconducting sample is shown. The larger concentration of field lines near the equator is a result of the demagnetisation factor.

## References


- L. D. Landau and E. M. Lifshitz (1977). Quantum Mechanics- Non-relativistic Theory. Butterworth-Heinemann. pp. 455–458. ISBN 0-7506-3539-8
- Drude, Paul (1900). "Zur Elektronentheorie der Metalle; II. Teil. Galvanomagnetische und thermomagnetische Effecte". Annalen der Physik. 308 (11): 369. Bibcode:1900AnP...308..369D. doi:10.1002/andp.19003081102
- Drakos, Nikos; Moore, Ross; Young, Peter (2002). "Landau diamagnetism". Electrons in a magnetic field. Retrieved 27 November 2012
- Poole, Jr., Charles P. (2007). Superconductivity (2nd ed.). Amsterdam: Academic Press. p. 23. ISBN 9780080550480
- Brady, R. M. (1982). "Correction to the Formula for the London Moment of a Rotating Superconductor". Journal of Low Temperature Physics. 49 (1): 1–17. Bibcode:1982JLTP...49....1B. doi:10.1007/bf00681758
- Herman, Robert; Prigogine, Ilya (April 1979). "A Two-Fluid Approach to Town Traffic" (PDF). Science. 204 (4389): 148–151. PMID 17738075. doi:10.1126/science.204.4389.148
- Jackson, Roland (21 July 2014). "John Tyndall and the Early History of Diamagnetism". Annals of Science: 4. doi:10.1080/00033790.2014.929743. Retrieved 28 October 2014
- Kittel, Charles (1986). Introduction to Solid State Physics (6th ed.). John Wiley & Sons. pp. 299–302. ISBN 0-471-87474-4
- Tate, J.; et al. (1990). "Determination of the Cooper-pair mass in niobium". Physical Review B. 42 (13): 7885. Bibcode:1990PhRvB..42.7885T. doi:10.1103/PhysRevB.42.7885

3

# Ginzburg-Landau Theory: An Overview

The Ginzburg-Landau theory considers the superconducting transition as one of the second order phase transitions. The amplitude ψ is zero in the normal phase above a superconducting transition temperature Tc. It is nonzero below a phase transition into a superconducting state. Superconductivity is best understood in confluence with the major topics listed in the following chapter.

## Ginzburg–Landau Theory

In physics, Ginzburg–Landau theory, often called Landau–Ginzburg theory, named after Vitaly Lazarevich Ginzburg and Lev Landau, is a mathematical physical theory used to describe superconductivity. In its initial form, it was postulated as a phenomenological model which could describe type-I superconductors without examining their microscopic properties. Later, a version of Ginzburg–Landau theory was derived from the Bardeen–Cooper–Schrieffer microscopic theory by Lev Gor'kov, thus showing that it also appears in some limit of microscopic theory and giving microscopic interpretation of all its parameters.

Based on Landau's previously-established theory of second-order phase transitions, Ginzburg and Landau argued that the free energy, $F$, of a superconductor near the superconducting transition can be expressed in terms of a complex order parameter field, $\psi$, which is nonzero below a phase transition into a superconducting state and is related to the density of the superconducting component, although no direct interpretation of this parameter was given in the original paper. Assuming smallness of $|\psi|$ and smallness of its gradients, the free energy has the form of a field theory.

$$F = F_n + \alpha |\psi|^2 + \frac{\beta}{2}|\psi|^4 + \frac{1}{2m}\left|(-i\hbar\nabla - 2e\mathrm{A})\psi\right|^2 + \frac{|\mathrm{B}|^2}{2\mu_0}$$

where $F_n$ is the free energy in the normal phase, $\alpha$ and $\beta$ in the initial argument were treated as phenomenological parameters, $m$ is an effective mass, $e$ is the charge of an electron, A is the magnetic vector potential, and $\mathrm{B} = \nabla \times \mathrm{A}$ is the magnetic field. By minimizing the free energy with respect to variations in the order parameter and the vector potential, one arrives at the Ginzburg–Landau equations

$$\alpha\psi + \beta|\psi|^2\psi + \frac{1}{2m}(-i\hbar\nabla - 2e\mathrm{A})^2\psi = 0$$

$$\nabla \times \mathrm{B} = \mu_0 \mathrm{j} \;;\; \mathrm{j} = \frac{2e}{m}\mathrm{Re}\left\{\psi^*(-i\hbar\nabla - 2e\mathrm{A})\psi\right\}$$

where j denotes the dissipation-less electric current density and *Re* the *real part*. The first equation — which bears some similarities to the time-independent Schrödinger equation, but is principally different due to a nonlinear term — determines the order parameter, $\psi$. The second equation then provides the superconducting current.

## Simple Interpretation

Consider a homogeneous superconductor where there is no superconducting current and the equation for $\psi$ simplifies to:

$$\alpha\psi + \beta|\psi|^2\psi = 0.$$

This equation has a trivial solution: $\psi = 0$. This corresponds to the normal state of the superconductor, that is for temperatures above the superconducting transition temperature, $T > T_c$.

Below the superconducting transition temperature, the above equation is expected to have a non-trivial solution (that is $\psi \neq 0$). Under this assumption the equation above can be rearranged into:

$$|\psi|^2 = -\frac{\alpha}{\beta}.$$

When the right hand side of this equation is positive, there is a nonzero solution for $\psi$ (remember that the magnitude of a complex number can be positive or zero). This can be achieved by assuming the following temperature dependence of $\alpha$: $\alpha(T) = \alpha_0\,(T - T_c)$ with $\alpha_0/\beta > 0$:

- Above the superconducting transition temperature, $T > T_c$, the expression $\alpha(T)/\beta$ is positive and the right hand side of the equation above is negative. The magnitude of a complex number must be a non-negative number, so only $\psi = 0$ solves the Ginzburg–Landau equation.
- Below the superconducting transition temperature, $T < T_c$, the right hand side of the equation above is positive and there is a non-trivial solution for $\psi$. Furthermore,

  $$|\psi|^2 = -\frac{\alpha_0(T - T_c)}{\beta},$$

  that is $\psi$ approaches zero as $T$ gets closer to $T_c$ from below. Such a behaviour is typical for a second order phase transition.

In Ginzburg–Landau theory the electrons that contribute to superconductivity were proposed to form a superfluid. In this interpretation, $|\psi|^2$ indicates the fraction of electrons that have condensed into a superfluid.

## Coherence Length and Penetration Depth

The Ginzburg–Landau equations predicted two new characteristic lengths in a superconductor which was termed coherence length, ξ. For T > $T_c$ (normal phase), it is given by

$$\xi = \sqrt{\frac{\hbar^2}{2m|\alpha|}}.$$

while for $T < T_c$ (superconducting phase), where it is more relevant, it is given by

$$\xi = \sqrt{\frac{\hbar^2}{4m|\alpha|}}.$$

It sets the exponential law according to which small perturbations of density of superconducting electrons recover their equilibrium value $\psi_0$. Thus this theory characterized all superconductors by two length scales. The second one is the penetration depth, λ. It was previously introduced by the London brothers in their London theory. Expressed in terms of the parameters of Ginzburg–Landau model it is

$$\lambda = \sqrt{\frac{m}{4\mu_0 e^2 \psi_0^2}},$$

where $\psi_0$ is the equilibrium value of the order parameter in the absence of an electromagnetic field. The penetration depth sets the exponential law according to which an external magnetic field decays inside the superconductor.

The original idea on the parameter "k" belongs to Landau. The ratio $\kappa = \lambda/\xi$ is presently known as the Ginzburg–Landau parameter. It has been proposed by Landau that Type I superconductors are those with $0 < \kappa < 1/\sqrt{2}$, and Type II superconductors those with $\kappa > 1/\sqrt{2}$.

The exponential decay of the magnetic field is equivalent with the Higgs mechanism in high-energy physics.

## Fluctuations in the Ginzburg–Landau model

Taking into account fluctuations. For Type II superconductors, the phase transition from the normal state is of second order, as demonstrated by Dasgupta and Halperin. While for Type I superconductors it is of first order as demonstrated by Halperin, Lubensky and Ma.

## Classification of Superconductors Based on Ginzburg–Landau Theory

In the original paper Ginzburg and Landau observed the existence of two types of superconductors depending on the energy of the interface between the normal and superconducting states. The Meissner state breaks down when the applied magnetic field is too large. Superconductors can be divided into two classes according to how this breakdown occurs. In Type I superconductors, superconductivity is abruptly destroyed when the strength of the applied field rises above a critical value $H_c$. Depending on the geometry of the sample, one may obtain an intermediate state consisting of a baroque pattern of regions of normal material carrying a magnetic field mixed with regions of superconducting material containing no field. In Type II superconductors, raising the applied field past a critical value $H_{c1}$ leads to a mixed state (also known as the vortex state) in which an increasing amount of magnetic flux penetrates the material, but there remains no resistance to the flow of electric current as long as the current is not too large. At a second critical field strength $H_{c2}$, superconductivity is destroyed. The mixed state is actually caused by vortices in the electronic superfluid, sometimes called fluxons because the flux carried by these vortices is quantized. Most pure elemental superconductors, except niobium and carbon nanotubes, are Type I, while almost all impure and compound superconductors are Type II.

The most important finding from Ginzburg–Landau theory was made by Alexei Abrikosov in 1957. He used Ginzburg–Landau theory to explain experiments on superconducting alloys and thin films. He found that in a type-II superconductor in a high magnetic field, the field penetrates in a triangular lattice of quantized tubes of flux vortices.

## Landau–Ginzburg Theories in String Theory

In particle physics, any quantum field theory with a unique classical vacuum state and a potential energy with a degenerate critical point is called a Landau–Ginzburg theory. The generalization to $N = (2,2)$ supersymmetric theories in 2 spacetime dimensions was proposed by Cumrun Vafa and Nicholas Warner in the November 1988 article Catastrophes and the Classification of Conformal Theories, in this generalization one imposes that the superpotential possess a degenerate critical point. The same month, together with Brian Greene they argued that these theories are related by a renormalization group flow to sigma models on Calabi–Yau manifolds in the paper Calabi–Yau Manifolds and Renormalization Group Flows. In his 1993 paper Phases of $N = 2$ theories in two-dimensions, Edward Witten argued that Landau–Ginzburg theories and sigma models on Calabi–Yau manifolds are different phases of the same theory. A construction of such a duality was given by relating the Gromov–Witten theory of Calabi–Yau orbifolds to FJRW theory an analogous Landau–Ginzburg "FJRW" theory in The Witten Equation, Mirror Symmetry and Quantum Singularity Theory. Witten's sigma models were later used to describe the low energy dynamics of 4-dimensional gauge theories with monopoles as well as brane constructions. Gaiotto, Gukov & Seiberg (2013).

The Ginzburg-Landau (GL) theory while not being a microscopic theory (but rather a macroscopic theory) provides a physically intuitive picture of the properties of superconductors. Broadly speaking, it starts by proposing a "wavefunction" $\psi(r)$ as a complex order parameter, characteristic of the superconducting state. Consequently, $|\psi(r)|^2$ is considered as the density of superconducting electrons $n_s(r)$. Next, it is assumed that the free-energy density of a superconductor can be expanded in powers of the supercarrier density as also in powers of $|\nabla\psi|^2$. After doing so, the free-energy density is minimized with respect to the order parameter to finally obtain the energy stabilisation of the superconducting state compared to the normal state. Note that the mechanism which drives superconductivity is not addressed by this theory but rather a practical approach is taken whereby given the fact that superconductivity exists, equations are provided to model their properties and in the process, new length scales such as the penetration depth and coherence length emerge. Functional description in different limiting cases is obtained. Further, based on energetics arguments, the existence of Type II superconductors can be hypothesized. The GL theory was proposed in 1950 (after Landau had given a general theory of second order phase transitions) which was before the BCS theory and in 1959, after the publication of the BCS theory, Gor'kov showed that the GL theory could in fact be rigorously obtained as a limiting case of the microscopic BCS theory.

## Free Energy Formulation

We now introduce the GL free energy density and then carry out a derivation of the GL equations followed by their application to specific geometries.

$$f_s = f_{n0} + \alpha |\psi|^2 + \frac{\beta}{2}|\psi|^4 + \frac{1}{2m^*}|(p - \frac{qA}{c})\psi|^2 + \frac{h^2}{8\pi}$$

Let us now elaborate on what the various terms in the RHS above are. Here, $f_{n0}$ is the (Helmholtz) free energy density in the absence of an applied magnetic field. Since the theory is considered to be applicable at temperatures close to the transition (where $n_s(r)$ is expected to be small), higher order terms in $|\psi(r)|^2$ and $|\nabla\psi|^2$ are neglected. One might ask as to why can we not expand in powers of the complex order parameter $\psi(r)$ as opposed to the square of its magnitude? The answer to this is that since the free energy is a real quantity, we need the RHS to be real as well. In that case, why can we not expand in powers of $|\psi|$ which is real and would then give a cubic term? The answer to this is that $|\psi|^3$ is not analytic at the origin and hence not admissible (derivative is different as one approaches the origin from different directions). The kinetic energy term takes into account the effect of applied external magnetic fields (here $\vec{p}$ is the operator $-i\hbar\vec{\nabla}$ and $(\vec{p} - \frac{q\vec{A}}{c})$ is the operator for the kinetic momentum; $(p_{kin} + \frac{qA}{c})$ is the canonical momentum). The mass and charge have been taken as $m^*$ and $q$ since at this point a connection with the electronic (or of pairs) mass and charge is not evident. The total kinetic energy would be given by

$$\int d^3r \frac{1}{2m^*}\psi^*(\vec{p} - \frac{q\vec{A}}{c})^2\psi$$

Integrating this by parts, the first term will vanish at the extrema since the wave function should be well behaved. The second term gives

$$\int dr \frac{1}{2m^*}|(\vec{p} - \frac{q\vec{A}}{c})\psi|^2$$

The integrand is then the energy density and this explains the electromagnetic term in the GL energy density. The last term is simply the magnetic self energy of the normal phase while $f_{n0}$ is free energy of the normal phase in zero applied field.

## Free Energy in the Absence of Field, Currents and Gradients

Let's us first consider the simplest situation where there are no magnetic fields nor are there any gradients of the wave function (this would correspond to a region well inside a superconductor). Here the free energy density reduces to

$$f_s = f_n + \alpha|\psi|^2 + \frac{\beta}{2}|\psi|^4$$

It is evident that both the coefficients $\alpha$ and $\beta$ can not be positive since that would not give rise to an energy lowering for finite $|\psi|^2$ and would in fact describe the normal state. The only physically meaningful possibility is for $\alpha$ to be negative and $\beta$ to be positive (either (i) the inverse or (ii) both having a negative sign would give lowest free energy for arbitrarily high values of $|\psi|^2$ and hence are not admissible). For a negative $\alpha$, taking the derivative of $f_s - f_n$ with respect to

$|\psi|^2$, we get

$$|\psi|^2 = -\frac{\alpha}{\beta}$$

The schematic variation of the free energy stabilisation as a function of $|\psi|^2$ is shown in figure for both positive and negative $\alpha$.

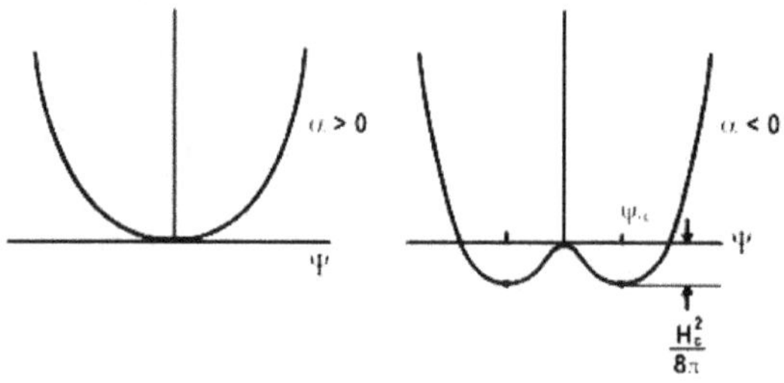


Characteristic length scale for superconducting wavefunction variation.

The above value of $|\psi|^2$ is also referred to as $|\psi_\infty|^2$ since we considered a location infinitely deep inside the superconductor. This then gives us $f_s - f_n = -\frac{\alpha^2}{2\beta}$. We can write this energy lowering in magnetic field units as equal to $-\frac{H_c^2}{8\pi}$ and define the thermodynamic critical field $H_c$ in this manner. In the above description, then, the change in the free energy curve in going from the normal state to the superconducting state must occur through a change in the coefficients $\alpha$ and $\beta$ with temperature. Since $\alpha$ changes sign at $T_c$, we can model the temperature dependence of $\alpha$ near $T_c$ using a Taylor series expansion $\alpha(T) = \alpha'(\frac{T}{T_c} - 1)$ where $\alpha'$ is the derivative of $\alpha$ with temperature and $\alpha' > 0$.The coefficient $\beta$ on the other hand remains positive across the transition. We now write the macroscopic wavefunction in terms of the magnitude and a phase factor as $|\psi|e^{i\varphi}$. The kinetic energy term in the GL equation can now be written as

$$\frac{1}{2m^*}|(\vec{p} - \frac{q\vec{A}}{c})\psi|^2 = \frac{1}{2m^*}|(-i\hbar\vec{\nabla} - \frac{q\vec{A}}{c})|\psi|e^{i\varphi}|^2$$

We now expand this by using the momentum operator for $p$. The result of applying the momentum operator on the macroscopic wavefunction is as follows

$$-i\hbar\vec{\nabla}(|\psi|e^{i\varphi}) = -i\hbar e^{i\varphi}\vec{\nabla}(|\psi|) + -i\hbar|\psi|e^{i\varphi}i\vec{\nabla}\varphi = -i\hbar e^{i\varphi}\vec{\nabla}(|\psi|) + \hbar|\psi|e^{i\varphi}\vec{\nabla}\varphi.$$

Therefore,

$$|(-i\hbar\vec{\nabla} - \frac{q\vec{A}}{c})|\psi|e^{i\varphi}|^2 = [-i\hbar e^{i\varphi}\vec{\nabla}(|\psi|) + (\hbar\vec{\nabla}\varphi - \frac{q\vec{A}}{c})|\psi|e^{i\varphi}][i\hbar e^{-i\varphi}\vec{\nabla}(|\psi|)$$
$$+\hbar\vec{\nabla}\varphi - \frac{q\vec{A}}{c})|\psi|e^{-i\varphi}].$$

## Determination of Coefficients a and b

The cross terms in the product above cancel out and some simple algebra yields for the energy $\frac{1}{2m^*}(\hbar^2\vec{\nabla}(|\psi|)^2+(\hbar\vec{\nabla}\varphi-\frac{q\vec{A}}{c})^2|\psi|^2)$. The first term gives the kinetic energy from the regions where there is a variation of the magnitude of the wavefunction such as at the boundary or in a domain wall. The second term involves the gradient of the phase and is associated with the supercurrent. This can be seen as follows. The quantum mechanical probability current density for a charged paricle in a magnetic field is given by

$$\frac{\hbar}{2m^*i}(\psi^*\vec{\nabla}\psi-\psi\vec{\nabla}\psi^*)-\frac{q}{m^*c}\vec{A}|\psi|^2=\frac{1}{m^*}(\hbar\vec{\nabla}\varphi-\frac{q\vec{A}}{c})|\psi|^2$$

Note that to get the electrical current density, one has to multiply this by the charge. Therefore the last term in the kinetic energy is $\frac{J^2m^*}{2q^2|\psi|^2}$. We know that the vector potenial is not unique and we can add the gradient of a scalar and still get the same magnetic field. We choose a gauge here (which is called the London Gauge) which cancels the $\hbar\vec{\nabla}\varphi$ term. Then the kinetic energy term (in the absence of a gradient of the magnitude of the wavefunction) becomes $\frac{q^2A^2|\psi|^2}{2m^*c^2}=\frac{A^2}{8\pi}\frac{4\pi q^2|\psi|^2}{mc^2}$. We have earlier seen that in the London model $\lambda_L^2=\frac{mc^2}{4\pi ne^2}$. The kinetic energy term now has the identification $\frac{A^2}{8\pi\lambda_{GL}^2}$ where $|\psi|^2$ is proportional to the superconducting carrier density, $q$ is the effective charge and $m^*$ is the effective mass. While it was not known at the time the theory was formulated as to what the effective mass and charge were (they were thought to be the bare electron mass and charge, respectively), it is now clear that due to the pairing of electrons, $m^*=2m$ and $q^*=2q$. Consequently, the density of superconducting pairs $n_s^*$ is half the density of electrons $n_s$. As a result, the expression for the GL penetration depth is the same as the London penetration depth.

$$\alpha=-\frac{H_c^2}{4\pi n_s}=-\frac{q^2}{mc^2}H_c^2(T)\lambda_{GL}^2(T)$$

We can now try to give physical meaning to the coefficients $\alpha$ and $\beta$. Since $\frac{H_c^2}{8\pi}=\frac{\alpha^2}{2\beta}$ and $\beta=-\frac{\alpha}{n_s}$,

we have

$$\alpha=-\frac{H_c^2}{4\pi n_s}=-\frac{q^2}{mc^2}H_c^2(T)\lambda_{GL}^2(T)$$

$$\beta=\frac{4\pi q^4}{m^2c^4}H_c^2(T)\lambda_{GL}^4(T).$$

Of course, $|\psi_\infty|^2 = n_s^* = \frac{mc^2}{4\pi q^2 \lambda_{GL}^2(T)}$. Here, $H_c(T)$ and $\lambda_{GL}(T)$ are experimentally measurable quantities. Using the empirical temperature dependence for the critical field and the penetration depth i.e., $H_c(T) \propto (1-(\frac{T}{T_c})^2)$ and $\frac{1}{(\lambda_{GL}(T))^2} \propto (1-(\frac{T}{T_c})^4)$, we obtain the temperature dependence of $\alpha$ and $\beta$.

$$\alpha(T) \propto \frac{(1-(\frac{T}{T_c})^2)^2}{(1-(\frac{T}{T_c})^4)} = \frac{1-(\frac{T}{T_c})^2}{1+(\frac{T}{T_c})^2} \simeq 1-\frac{T}{T_c}$$

$$\beta(T) \propto \frac{(1-(\frac{T}{T_c})^2)^2}{(1-(\frac{T}{T_c})^4)^2} = \frac{1}{(1+(\frac{T}{T_c})^2)^2} \simeq \text{constant}$$

Here we have expanded in powers of $1-\frac{T}{T_c}$ and kept only the leading terms. In all the above treatment, we have ignored any spatial variation of the order parameter inside the superconductor.

## GL Equations in the Presence of Field, Currents and Gradients

We have presently solved for a situation where there were no fields, currents or gradients. In case these are present we need to minimise the free energy containing the contribution of these terms i.e. Eqn. 1.

$$F = \int f_s d^3x = \int (f_{n0} + \alpha|\psi|^2 + \frac{\beta}{2}|\psi|^4 + \frac{1}{2m^*}|(-i\hbar\vec{\nabla} - \frac{q\vec{A}}{c})\psi|^2 + \frac{h^2}{8\pi}) d^3x$$

This is really a free energy functional which is a scalar number but it depends on $\psi$ (and $\psi^*$ and $A$) at all coordinate points in the system. The free energy has to be minimised wrt variations of $\psi, \psi^*$ and $A$.

This means the following.

$$dF = \int [\frac{\partial f_s}{\partial \psi} d\psi + \frac{\partial f_s}{\partial \psi^*} d\psi^* + \frac{\partial f_s}{\partial A} dA] d^3x = 0$$

To evaluate the above, let us write the free energy density in an expanded form

$$f_s - f_{n0} = \alpha|\psi|^2 + \frac{\beta}{2}|\psi|^4 + \frac{1}{2m^*}|(-i\hbar\vec{\nabla} - \frac{q\vec{A}}{c}\psi|^2 + \frac{h^2}{8\pi}$$

$$= \alpha\psi^*\psi + \frac{\beta}{2}\psi^2\psi^{*2} + \frac{1}{2m^*}[\hbar^2|\vec{\nabla}\psi|^2 + \frac{q^2\psi^*\psi A^2}{c^2} + \frac{i\hbar q A}{c}(\psi^*\vec{\nabla}\psi - \psi\vec{\nabla}\psi^*)] + \frac{(\vec{\nabla}\times\vec{A})^2}{8\pi}$$

Taking partial derivative with respect to $\psi^*$ gives

$$\int [\frac{\partial f_s}{\partial \psi^*} d\psi^*] d^3x = \int [\alpha\psi + \beta|\psi|^2 \psi + \frac{1}{2m^*}\frac{\partial}{\partial \psi^*}[\hbar^2 |\vec{\nabla}\psi|^2 + \frac{q^2\psi^*\psi A^2}{c^2} + \frac{i\hbar q A}{c}(\psi^*\vec{\nabla}\psi - \psi\vec{\nabla}\psi^*)]] d\psi^* d^3x$$

Note that $\int \frac{1}{2m^*}\frac{\partial}{\partial \psi^*}[\hbar^2 |\vec{\nabla}\psi|^2] d\psi^* d^3x$ can be integrated by parts (write $|\vec{\nabla}\psi|^2 = \vec{\nabla}\psi^* \cdot \vec{\nabla}\psi$ and $\int \vec{\nabla}u \cdot \vec{v} d^3x = s\int u\vec{v}\cdot dS - \int u\vec{\nabla}\cdot\vec{v} d^3x$) to give $\frac{-\hbar^2}{2m^*}\int \nabla^2\psi d\psi^* d^3x$ + a surface integral (let us call this S1). Similarly

$$\int \frac{1}{2m^*}\frac{\partial}{\partial \psi^*}[\frac{q^2\psi^*\psi A^2}{c^2}] d\psi^* d^3x = \int \frac{q^2\psi A^2}{2m^* c^2} d\psi^* d^3x$$

and

$$\int \frac{1}{2m^*}\frac{\partial}{\partial \psi^*}[\frac{i\hbar q A\cdot}{c}(\psi^*\vec{\nabla}\psi)] d\psi^* d^3x = \int \frac{i\hbar q A\cdot}{2m^* c}(\vec{\nabla}\psi) d\psi^* d^3x$$

Further $\int \frac{1}{2m^*}\frac{\partial}{\partial \psi^*}[\frac{i\hbar q\vec{A}\cdot}{c}(-\psi\vec{\nabla}\psi^*)] d\psi^* d^3x$ can be integrated by parts. This integration yields the above volume integral + a surface integral (let us call this S2). Combine the two surface integrals S1 and S2, and impose the following boundary condition to make them zero. (normal component of $(-i\hbar\vec{\nabla} - \frac{q\vec{A}}{c})\psi$ on the bounding surface is zero). Putting the integrand corresponding to $\frac{\partial f_s}{\partial \psi^*} = 0$, we get

$$\alpha\psi + \beta|\psi|^2\psi + \frac{1}{2m^*}|(-i\hbar\vec{\nabla} - \frac{q\vec{A}}{c})^2|\psi = 0$$

This is called the first GL equation. Minimising the energy with respect to $\psi$ just yields the complex conjugate of this equation. Next we need to minimise the free energy wrt $A$. For this we need to consider only the terms in the free energy which depend on $A$. These are given below

$$\frac{1}{2m^*}[\frac{q^2\psi^*\psi A^2}{c^2} + \frac{i\hbar q A}{c}(\psi^*\vec{\nabla}\psi - \psi\vec{\nabla}\psi^*)] + \frac{(\vec{\nabla}\times\vec{A})^2}{8\pi}$$

The variation of the above terms due to a variation $\delta A$ to $A$ is $[\frac{q^2|\psi|^2 A}{m^* c^2} + \frac{i\hbar q}{2m^* c}(\psi^*\vec{\nabla}\psi - \psi\vec{\nabla}\psi^*)]\delta A$.

Now we need to do the same for the $\frac{(\vec{\nabla}\times\vec{A})^2}{8\pi}$ term.

$$(\vec{\nabla}\times(\overrightarrow{A+\delta A}))^2$$

$$=\vec{\nabla}\times(\overrightarrow{A+\delta A})\cdot\vec{\nabla}\times(\overrightarrow{A+\delta A})$$

$$=(\vec{\nabla}\times\vec{A})^2+(\vec{\nabla}\times\overrightarrow{\delta A})^2+2(\vec{\nabla}\times\vec{A}).(\vec{\nabla}\times\overrightarrow{\delta A})$$

The change in energy due to variation of $A$ (keeping only the term linear in $\delta A$) is $\frac{(\vec{\nabla}\times\vec{A})\cdot(\vec{\nabla}\times\overrightarrow{\delta A})}{4\pi}$. Now use the vector identity

$$\vec{\nabla}\cdot(\vec{C}\times\vec{D})=\vec{D}.\vec{\nabla}\times\vec{C}-\vec{C}\cdot\vec{\nabla}\times\vec{D}\Rightarrow\vec{D}.\vec{\nabla}\times\vec{C}=\vec{\nabla}\cdot\left(\vec{C}\times\vec{D}\right)+\vec{C}\cdot\vec{\nabla}\times\vec{D}\cdot$$

$$\frac{(\vec{\nabla}\times\vec{A}).(\vec{\nabla}\times\overrightarrow{\delta A})}{4\pi}=\frac{\vec{\nabla}.(\overrightarrow{\delta A}\times(\vec{\nabla}\times\vec{A}))}{4\pi}+\frac{\overrightarrow{\delta A}.\vec{\nabla}\times(\vec{\nabla}\times\vec{A})}{4\pi}$$

The two terms in the above equation will appear in the volume integral of the energy density. The first term can be converted to the surface integral $\frac{1}{4\pi}\oint(\overrightarrow{\delta A}\times(\vec{\nabla}\times\vec{A}))\cdot\overrightarrow{dS}$ using the divergence theorem. This will be zero since $\overrightarrow{\delta A}\times(\vec{\nabla}\times\vec{A})$ is directed perpendicular to $\overrightarrow{dS}$. The remaining term is $\frac{\overrightarrow{\delta A}\cdot\vec{\nabla}\times(\vec{\nabla}\times\vec{A})}{4\pi}$ which is the same as $\frac{\vec{J}\cdot\overrightarrow{\delta A}}{c}$. Combining this with $[\frac{q^2|\psi|^2\vec{A}}{m^*c^2}+\frac{i\hbar q}{2m^*c}(\psi^*\vec{\nabla}\psi-\psi\vec{\nabla}\psi^*)]\delta A$ we say that the volume integral must be zero for the energy to be minimum with respect to arbitrary variations of $A$. Therefore, the integrand must be zero at all points in the volume. Therefore we get the supercurrent density

$$\vec{J}=-\frac{q^2|\psi|^2\vec{A}}{m^*c}-\frac{i\hbar q}{2m^*}(\psi^*\vec{\nabla}\psi-\psi\vec{\nabla}\psi^*)$$

The two GL equations are now summarised below:

$$\alpha\psi+\beta|\psi|^2\psi+\frac{1}{2m^*}(-i\hbar\vec{\nabla}-\frac{q\vec{A}}{c})^2\psi=0$$

$$\vec{J}=\frac{q|\psi|^2}{m^*}(\hbar\vec{\nabla}\varphi-\frac{q\vec{A}}{c})$$

The first equation is very similar to the Schroedinger equation (except for the $\beta|\psi|^2\psi$ term). The non-linear term tends to favour wave functions which have small or no spatial variations. In the case of the appearance of an interface (such as between a superconductor and a normal metal), clearly the wavefunction (or the order parameter) will become zero at the surface while deep inside the superconductor it will have its maximum value.

## Superconducting Coherence Length

In superconductivity, the superconducting coherence length, usually denoted as $\xi$ is the characteristic exponent of the variations of the density of superconducting component.

In some special limiting cases, for example in the weak-coupling BCS theory it is related to characteristic Cooper pair size.

The superconducting coherence length is one of two parameters in the Ginzburg-Landau theory of superconductivity. It is given by:

$$\xi = \sqrt{\frac{\hbar^2}{2m|\alpha|}}$$

where $\alpha$ is a constant in the Ginzburg-Landau equation for $\psi$ with the form $\alpha_0(T-T_c)$. In BCS theory:

$$\xi_{BCS} = \frac{\hbar v_f}{\pi\Delta}$$

where $\hbar$ is the reduced Planck constant, $m$ is the mass of a Cooper pair (twice the electron mass), $v_f$ is the Fermi velocity, and $\Delta$ is the superconducting energy gap.

The ratio $\kappa = \lambda/\xi$, where $\lambda$ is the London penetration depth, is known as the Ginzburg–Landau parameter. Type-I superconductors are those with $0 < \kappa < 1/\sqrt{2}$, and type-II superconductors are those with $\kappa > 1/\sqrt{2}$.

For temperatures $T$ near the superconducting critical temperature $T_c$, $\xi(T) \propto (1-T/T_c)^{-1}$.

### GL Coherence Length

We will now see that the length scale over which the wavefunction recovers its full value is $\sqrt{\frac{\hbar^2}{2m^*|\alpha|}}$ and is called the GL coherence length. Considering the 1st GL equation for a one-dimensional case, in the absence of electromagnetic fields and normalising the equation to $\psi_\infty$, we get the following (where $f = \frac{\psi}{\psi_\infty}$ and $\psi_\infty^2 = -\frac{\alpha}{\beta}$)

$$\frac{\hbar^2}{2m^*\alpha}\frac{d^2 f}{dx^2} + f - f^3 = 0$$

Since $f$ is dimensionless, it is clear that $\xi(T)^2 = \frac{\hbar^2}{2m^*|\alpha(T)|}$ has dimensions of length squared. If we define $y = \frac{x}{\xi}$, the above equation can be written in the dimensionless form

$$\frac{d^2 f}{dy^2} + f - f^3 = 0$$

We try a solution $f(y) = \tanh(cy)$. This gives $2c^2 = 1$ so $f(y) = \tanh(\frac{y}{\sqrt{2}})$. That $\xi$ represents a lengthscale over which $\psi$ varies can be seen in another manner. While the above equation is non-linear, we can "linearise" the equation in the following way. Let $f(y) = 1 + g(y)$ where $g \ll 1$. Substitute this in the above equation and keep only the linear terms in $g$. This gives $g''(y) = 2g(y)$ which has the solution $g(y) \sim e^{\pm y\sqrt{2}}$. We have earlier written $\alpha$ in terms of the critical field and penetration depth. We now use that to express the coherence length in the following form

$$\xi(T) = \frac{\phi_0}{2\sqrt{2}\pi H_c(T)\lambda(T)}$$

where $\phi_0 = \frac{hc}{q} = \frac{hc}{2e}$ is the fluxoid quantum. The fluxoid quantum can also be written in terms of the London penetration depth and the Pippard coherence length using results from the BCS theory, more specifically, $\xi_0 = \hbar v_F / \pi\Delta(0)$ and $H_c^2(0)/8\pi = \frac{1}{2}N(0)\Delta^2(0)$. This leads to

$$\phi_0(T) = \sqrt{\frac{2}{3}}\pi^2\xi_0\lambda_L(0)H_c(0)$$

Therefore, the ratio of the GL coherence length to the Pippard coherence length can be written as

$$\frac{\xi(T)}{\xi_0} = \frac{\pi}{2\sqrt{3}}\frac{H_c(0)}{H_c(T)}\frac{\lambda_L(0)}{\lambda_{GL}(T)}$$

Note that a superconductor is said to be in the clean limit if the electronic mean free path $l$ in it is much less than the coherence length $\xi$. In case $l < \xi$, the superconductor is said to be in the dirty limit. Using the BCS expressions for the variation of the critical field and penetration depth neat $T_c$, we obtain the variation of the coherence length with temperature in the clean and dirty limits as follows

$$\xi(T) = 0.74\frac{\xi_0}{(1 - T/T_c)^{1/2}} \; pure$$

$$\xi(T) = 0.855\frac{(\xi_0 l)^{1/2}}{(1 - T/T_c)^{1/2}} \; \text{dirty}$$

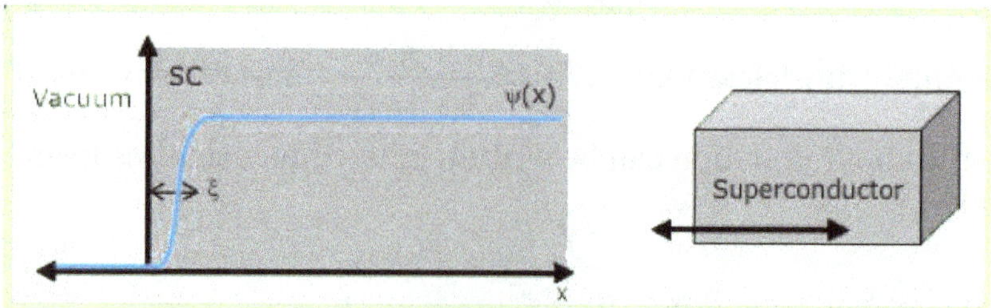


Characteristic length scale for superconducting wavefunction variation.

A parameter which characterises two classes of superconductors is $\kappa = \frac{\lambda}{\xi}$. Typical values of $\xi$ range from a few tens of $\AA$ (high- $T_c$ superconductors) to thousands of $\AA$ (elemental Al).

Accompanying figure for the schematic variation of the order parameter.

## Penetration of Magnetic Field in a Semi-infinite Slab in the GL Approach

Let us now consider the problem of penetration of a weak magnetic field inside a superconductor with a planar boundary ($yz$ – plane). Consider the magnetic field along the $z$ -axis. In this geometry, the field will penetrate the superconductor and decay as we go along the $x$ -direction. Let's choose a vector potential $A$ which has a non-zero $y$ -component $A_y$. We then have $H(x) = \frac{dA_y}{dx}$. Further, it is natural to assume that $\psi$ has only an $x$ -dependence. Let us redefine various quantities so that we can work with dimensionless quantities.

$$f = \frac{\psi}{\psi_\infty}$$

$$r' = \frac{r}{\lambda}$$

$$A' = \frac{A}{\sqrt{2}H_c\lambda}$$

Now define $\kappa = \frac{\lambda}{\xi}$. We then rewrite the GL equations dropping the primes.

$$(-\frac{i\vec{\nabla}}{\kappa} - \vec{A})^2 f - f + |f|^2 f = 0$$

$$\vec{\nabla} \times \vec{\nabla} \times \vec{A} = -\frac{i}{2K}(f^* \vec{\nabla} f - f \vec{\nabla} f^*) - |f|^2 \vec{A}$$

For the given case we get,

$$-\frac{1}{\kappa^2}\frac{d^2 f}{dx^2} - f + A^2 f + |f|^2 f = 0$$

Assuming that $\kappa \ll 1$

$$\frac{d^2 f}{dx^2} = 0$$

Further, at the boundary $\frac{df}{dx} = 0$. So $f = \text{constant}$ and $|f|^2 = 1$ From the second GL eqn., we get

$$\frac{d^2 A}{dx^2} - A = 0$$

A similar equation is obtained for $H$ and solving it (using the boundary condition at the surface) gives, in non-reduced units,

$$H = H_0 e^{-\frac{x}{\lambda}}$$

where $\lambda$ is the London penetration depth.

## Spatial Variation of the Order Parameter for a Semi-infinite Slab

We will now illustrate the solution for a situation which is not limited to small $\kappa$ which lead to non-constancy of $f$. For simplicity, consider a one-dimensional case where $f$ and $A$ depend on a single coordinate $x$ and $A$ is perpendicular to $x$. Also, we assume that $f$ is real. Therefore,

$$\frac{1}{\kappa^2}\frac{d^2 f}{dx^2} + (1 - A^2) f - f^3 = 0$$

Also,

$$\frac{d^2 A}{dx^2} - f^2 A = 0$$

Multiply the first equation by $\frac{df}{dx}$ and the second by $\frac{dA}{dx}$, add the two equations and integrate over $x$.

This leads to the following

$$\frac{1}{\kappa^2}(\frac{df}{dx})^2 + (\frac{dA}{dx})^2 + f^2(1 - A^2) - \frac{1}{2} f^4 = \text{constant}$$

Note, once again, that $f$ represents the normalised order parameter $x$ has been normalised by $\lambda$, and the vector potential has been normalised by $\sqrt{2}H_c\lambda$. Assuming that a field equal to the critical field is applied, let's apply the boundary conditions, (i) well into the interior of the superconductor $f = 1$ and $H = A = 0$ and (ii) well outside the superconductor $f = 0$ and the field in reduced units is $H = \frac{H_0}{H_c\sqrt{2}} = \frac{1}{\sqrt{2}}$. This fixes the constant in the above equation to $\frac{1}{2}$. For a solution to the above equation, let's consider the limiting case of $\kappa \ll 1$. Further consider $A$ and $H$ to be small in the above equation. This gives

$$\frac{df}{dx} = \frac{\kappa}{\sqrt{2}}(1 - f^2)$$

The solution to this equation satisfying the boundary condition (the region from $x = 0\, to\, x = \infty$ is the superconductor) is $f = \tanh(\frac{\kappa x}{\sqrt{2}})$.

## Type-II Superconductor

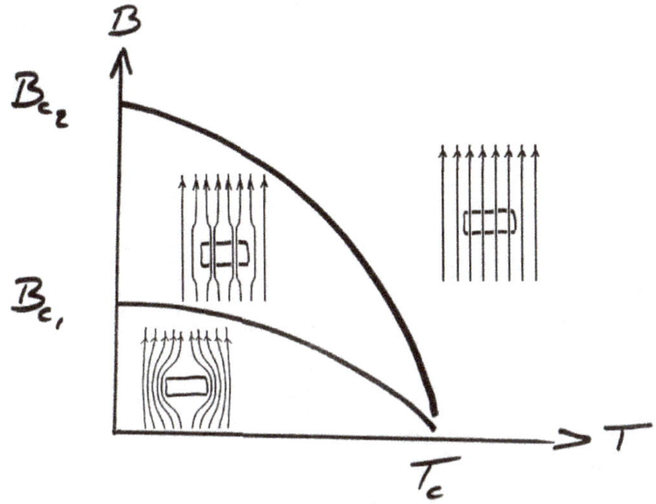


The B-T diagram of type-II superconductor.

In superconductivity, a type-II superconductor is characterized by the formation of magnetic vortices in an applied magnetic field. This occurs above a certain critical field strength $H_{c1}$. The vortex density increases with increasing field strength. At a higher critical field $H_{c2}$, superconductivity is destroyed. Type-II superconductors do not exhibit a complete Meissner effect.

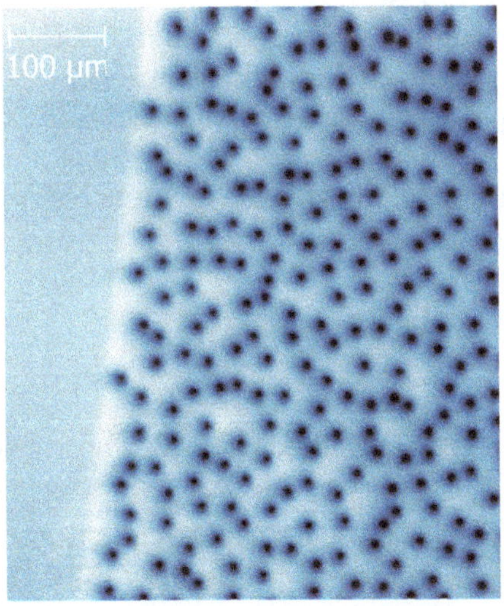


Vortices in a 200-nm-thick YBCO film imaged by scanning SQUID microscopy.

### History

In 1935, Rjabinin, Lev Shubnikov experimentally discovered the Type-II superconductors. In 1950, the theory of the two types of superconductors was further developed by Lev Landau and Vitaly Ginzburg in their paper on Ginzburg-Landau theory. In their argument, a type-I superconductor had positive free energy of the superconductor-normal metal boundary. Ginzburg and Landau pointed out the possibility of type-II superconductors that should form inhomogeneous state in strong magnetic fields. However, at that time, all known superconductors were type-I, and they commented that there was no experimental motivation to consider precise structure of type-II superconducting state. The theory for the behavior of the Type-II superconducting state in magnetic field was greatly improved by Alexei Alexeyevich Abrikosov, who was elaborating on the ideas by Lars Onsager and Richard Feynman of quantum vortices in superfluids. Quantum vortex solution in a superconductor is also very closely related to Fritz London's work on magnetic flux quantization in superconductors. The Nobel Prize in Physics was awarded for the theory of Type-II superconductivity in 2003.

## Vortex State

Ginzburg–Landau theory defines two parameters: The superconducting coherence length and the London magnetic field penetration depth. In a type-II superconductor, the coherence length is smaller than the penetration depth. This leads to negative energy of the interface between superconducting and normal phases. The existence of the negative interface energy was known since the mid-1930s from the early works by the London brothers. A negative interface energy suggests that the system should be unstable against maximizing the number of such interfaces, which was not observed in first experiments on superconductors, before the experiments of Shubnikov in 1936 where two critical fields were found. In 1952 observation of type-II superconductivity was also reported by Zavaritskii. As was later discussed by A. A. Abrikosov, these interfaces manifest as lines of magnetic flux passing through the material, turning a region of the superconductor normal. This normal region is separated from the rest of the superconductor by a circulating supercurrent. In analogy with fluid dynamics, the swirling supercurrent creates what is known as a *vortex*, or an Abrikosov vortex, after Alexei Alexeyevich Abrikosov. In the limit of very short coherence length the vortex solution is identical to London's fluxoid, where vortex core is approximated by sharp cutoff rather than gradually vanishing of superconducting condensate near the vortex center. Abrikosov found that the vortices arrange themselves into a regular array known as a *vortex lattice*.

In the extreme type-II limit, the problem of type-II superconductor in magnetic field is exactly equivalent to that of vortex state in rotating superfluid helium, which was discussed earlier by Richard Feynman in 1955.

## Flux Pinning

In the vortex state, a phenomenon known as flux pinning, where a superconductor is pinned in space above a magnet, becomes possible. This is not possible with type-I superconductors, since they cannot be penetrated by magnetic fields. Since the superconductor is pinned above the magnet away from any surfaces, there is the potential for a frictionless joint. The worth of flux pinning is seen through many implementations such as lifts, frictionless joints, and transportation. The thinner the superconducting layer, the stronger the pinning that occurs when exposed to magnetic fields.

## Materials

Type-II superconductors are usually made of metal alloys or complex oxide ceramics. All high temperature superconductors are type-II superconductors. While most elemental superconductors are type-I, niobium, vanadium, and technetium are elemental type-II superconductors. Boron-doped diamond and silicon are also type-II superconductors. Metal alloy superconductors also exhibit type-II behavior (*e.g.* niobium-titanium and niobium-tin).

Other type-II examples are the cuprate-perovskite ceramic materials which have achieved the highest superconducting critical temperatures. These include $La_{1.85}Ba_{0.15}CuO_4$, BSCCO, and YBCO (Yttrium-Barium-Copper-Oxide), which is famous as the first material to achieve superconductivity above the boiling point of liquid nitrogen (77 K). Due to strong vortex pinning, the cuprates are close to ideally hard superconductors.

## Important Uses

Strong superconducting electromagnets (used in MRI scanners, NMR machines, and particle accelerators) often use niobium-titanium or, for higher fields, niobium-tin.

## Stabilisation of Type-II superconductivity: Surface Energy Calculation

A calculation of the energy corresponding to the surface between superconducting and non-superconducting regions will now be performed. It will be shown that depending on the value of $\kappa$ either a Type-I or a Type-II superconductor will be stabilised. We have earlier written the free energy for a superconducting sample in the GL aproach. In the reduced units introduced above, this can be rewritten as

$$\int(f_s - f_{n0})d^3x = \int\frac{H_c^2}{4\pi}[-|f|^2 + \frac{1}{2}|f|^4 + |(-\frac{i\vec{\nabla}}{\kappa} - \vec{A})f|^2 + H^2]d^3x$$

$$= \int\frac{H_c^2}{4\pi}[-|f|^2 + \frac{1}{2}|f|^4 + \frac{i}{\kappa}(-\frac{i\vec{\nabla}}{\kappa} - \vec{A})f.\vec{\nabla}f^* + \frac{i\vec{A}.f^*\vec{\nabla}f}{\kappa} + A^2f^2 + H^2]d^3x$$

Then we integrate the term with $\vec{\nabla}f^*$ by parts. Here the following rules for integration by parts are useful.

$$\int\vec{\nabla}u\cdot\vec{v}d^3x = s\int u\vec{v}.d\vec{S} - \int u\vec{\nabla}\cdot\vec{v}d^3x$$

The surface integral is zero due to our boundary conditions. Then we get

$$= \int\frac{H_c^2}{4\pi}[-|f|^2 + \frac{1}{2}|f|^4 - \frac{i}{\kappa}f^*\vec{\nabla}\cdot(-\frac{i\vec{\nabla}f}{\kappa} - \vec{A}f) + \frac{i\vec{A}\cdot f^*\vec{\nabla}f}{\kappa} + A^2f^2 + H^2]d^3x$$

$$= \int\frac{H_c^2}{4\pi}[-|f|^2 + \frac{1}{2}|f|^4 - \frac{f^*\nabla^2 f}{\kappa^2} + \frac{i}{\kappa}f^*\vec{\nabla}\cdot(\vec{A}f) + \frac{i\vec{A}\cdot f^*\vec{\nabla}f}{\kappa} + A^2f^2 + H^2]d^3x$$

$$= \int\frac{H_c^2}{4\pi}[-|f|^2 + \frac{1}{2}|f|^4 - \frac{f^*\nabla^2 f}{\kappa^2} + \frac{i}{\kappa}f^*\vec{\nabla}\cdot(\vec{A}f) + \frac{i\vec{A}\cdot f^*\vec{\nabla}f}{\kappa} + A^2f^2 + H^2]d^3x$$

$$= \int\frac{H_c^2}{4\pi}[-|f|^2 + \frac{1}{2}|f|^4 - \frac{f^*\nabla^2 f}{\kappa^2} + \frac{2i}{\kappa}f^*\vec{A}\cdot\vec{\nabla}f + A^2f^2 + H^2]d^3x$$

$$= \int\frac{H_c^2}{4\pi}[-|f|^2 + \frac{1}{2}|f|^4 + f^*(-\frac{i\vec{\nabla}}{\kappa} - \vec{A})^2 f + H^2]d^3x$$

Using the first GL equation, this gives

$$\int \frac{H_c^2}{4\pi}[-|f|^2 + \frac{1}{2}|f|^4 + |f|^2 - |f|^4 + H^2]d^3x$$

$$= \int \frac{H_c^2}{4\pi}[H^2 - \frac{1}{2}|f|^4]d^3x$$

Now consider an actual case of a cylindrical conductor in which we look at the transition from the normal to the superconducting state under the influence of a field applied parallel to the axis of the cylinder. We have shown earlier that that the normal state free energy in a field $H_0$ is reduced from its zero-field value by $\frac{H_0^2}{8\pi}$. The free energy difference (in reduced units) between the superconducting state and the normal state, for a cylindrical sample in a magnetic field $H_0$ is ( $H$ is the field in the sample)

$$\int (f_{sH} - f_{nH})d^3x = \frac{H_c^2}{4\pi}\int [(H - H_0)^2 - \frac{1}{2}|f|^4]\mathrm{d}^3x$$

For the $f = \tanh(\frac{\kappa x}{\sqrt{2}})$ solution obtained earlier, which had an applied field equal to the critical field ( $H_0 = \frac{1}{\sqrt{2}}$) and the internal field was zero. The surface energy density (energy per unit area) is

$$\sigma_{ns} = \frac{H_c^2}{8\pi}\int (1 - f^4)dx$$

$$= \frac{H_c^2}{8\pi}\int (1 - \tanh^4(\frac{\kappa x}{\sqrt{2}}))dx$$

$$= \frac{H_c^2}{8\pi}\int \cosh^{-2}(\frac{\kappa x}{\sqrt{2}})(1 + \tanh^2(\frac{\kappa x}{\sqrt{2}}))dx$$

Here, the limits of the integral are from $0$ to $\infty$. The integration gives

$$= \frac{H_c^2}{8\pi}\frac{4\sqrt{2}}{3\kappa}$$

The energy per unit area will have dimensions of energy density times length. In conventional units, this is

$$= \frac{H_c^2}{8\pi}\frac{4\sqrt{2}\lambda}{3\kappa}$$

Remember that the above calculation was for $\kappa \ll 1$. For $\kappa \gtrsim 1$, the surface energy calculation has to be done numerically. It is clear from the free energy expression in a reduced field (i.e., in reduced units) $H_0 = \frac{1}{\sqrt{2}}$ that it will be zero when $H = \frac{dA}{dx} = \frac{1-f^2}{\sqrt{2}}$. This gives $\frac{d\psi}{dx} = -\frac{A\psi}{\sqrt{2}}$. We can substitute this in the equation

$$\frac{1}{\kappa^2}(\frac{df}{dx})^2 + (\frac{dA}{dx})^2 + f^2(1-A^2) - \frac{1}{2}f^4 = \frac{1}{2}$$

to obtain $\kappa = \frac{1}{\sqrt{2}}$. The energy is negative at higher $\kappa$ so the material prefers to have more surfaces in that case. This is the case of Type II superconductors where one has normal regions inside the superconductor.

## Surface Energy and "Type II"

In the former case, there is a net positive energy due to creation of a surface. This leads to Type-I superconductivity. In the latter case, there is a net negative surface energy. This means that the sample prefers to have "surfaces" where the order parameter goes to zero. This leads to creation of interfaces to lower the overall energy. Such superconductors where there is a penetration of magnetic field inside them (creating many interfaces with normal regions) are called Type-II superconductors.

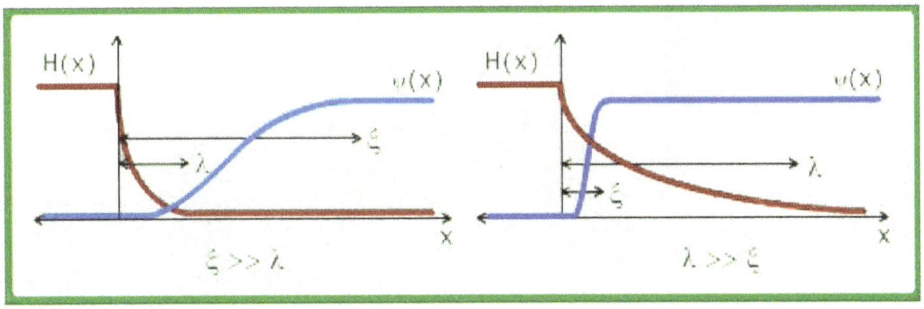


The variation of superconducting order parameter inside a superconductor in two extreme situations is shown.

These things are illustrated in the following two figures.

$$Surface\ Energy: \xi > \lambda$$

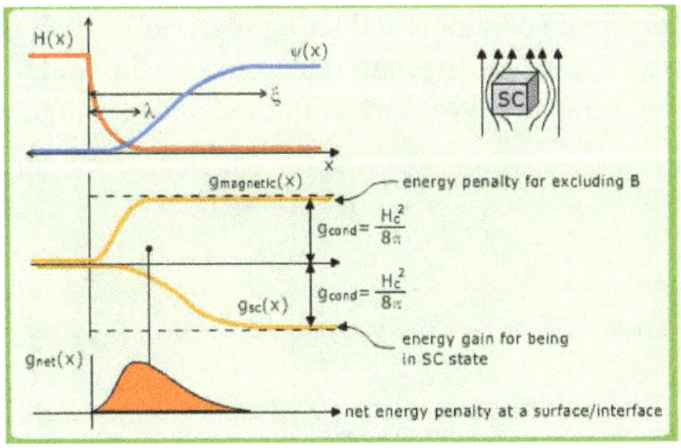

*Surface Energy*: $\xi < \lambda$

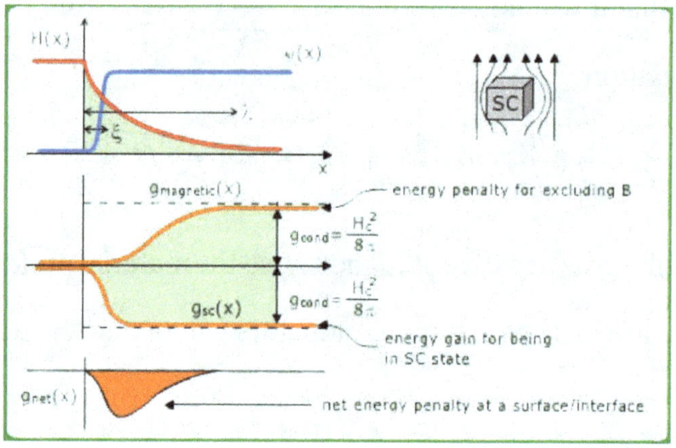


*Type II Superconductors* $(\xi < \lambda)$

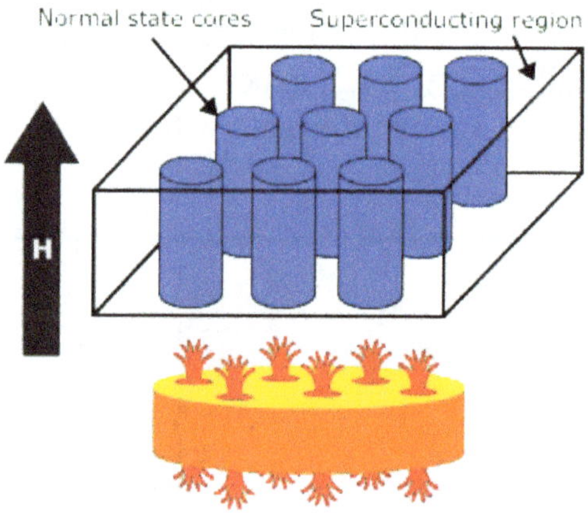


Since Type-II superconductors typically have a very high upper critical field, the field range in which they can be used is large and hence they find many practical applications. However, there are many superconductors which have $\kappa \ll 1$ (0.16 for Hg, 0.15 for Sn, and 0.026 for Al).

## Fluxoid Quantisation

Consider a hollow cylinder (made of superconducting material) which is placed in an axial magnetic field, less than the critical field for the material. Consider the wall thickness of the cylinder to be greater than the penetration depth. We start from the London equation

$$\frac{\partial \vec{J}}{\partial t} = \frac{nq^2}{m^*}\vec{E} = \frac{c^2}{4\pi\lambda^2}\vec{E}$$

which can be re-written as

$$\frac{4\pi\lambda^2}{c^2}\frac{\partial \vec{J}}{\partial t} = \vec{E}$$

We integrate this equation along a closed contour lying entirely within the superconductor (and not within the surface). This gives

$$\frac{4\pi\lambda^2}{c^2}\frac{\partial}{\partial t}\oint \vec{J}.d\vec{l} = \oint \vec{E}.d\vec{l} = \int \vec{\nabla}\times\vec{E}\cdot d\vec{S}$$

Using Maxwell equations this gives

$$\frac{4\pi\lambda^2}{c^2}\frac{\partial}{\partial t}\oint \vec{J}.d\vec{l} = -\frac{1}{c}\int \frac{\partial \vec{B}}{\partial t}\cdot d\vec{S}$$

or

$$\frac{4\pi\lambda^2}{c^2}\oint \vec{J}\cdot d\vec{l} + \frac{1}{c}\int \vec{B}\cdot d\vec{S} = \text{constant}$$

Now, if the contour that we have chosen is well inside the superconductor and not in the surface layer, the current $J = 0$. This implies that the flux threading the superconducting loop does not vary with time. In the following we will show (using GL equations) that the trapped flux cannot be arbitrary but is rather an integral multiple of a fundamental quantity: the flux quantum. In general, the current density in the superconductor is given by the second GL equation

$$\vec{J} = \frac{q|\psi|^2}{m^*}(\hbar\vec{\nabla}\varphi - \frac{q\vec{A}}{c})$$

We re-write this is as

$$\frac{\vec{J}}{|\psi|^2} = \frac{q\hbar}{m^*}(\vec{\nabla}\varphi - \frac{q\vec{A}}{\hbar c})$$

and integrate along the contour as before

$$\oint \frac{\vec{J}}{|\psi|^2}\cdot d\vec{l} = \frac{q\hbar}{m^*}\oint(\vec{\nabla}\varphi - \frac{q\vec{A}}{\hbar c})\cdot d\vec{l}$$

Once again, if the contour is chosen such that $J = 0$ then

$$\frac{\hbar c}{q}\oint \vec{\nabla}\varphi.d\vec{l} = \oint \vec{A}\cdot d\vec{l} = \int \vec{B}\cdot d\vec{S} = \phi$$

Since the superconducting wave function must be single valued at any point, the phase can change by an integral multiple of $2\pi$ on coming back to a point after traversing the contour. Therefore, the integral on the left hand side should be an integral multiple of $2\pi$. So $\phi = n\phi_0$ where $\phi_0 = \frac{hc}{2e} = 2.07\times10^{-7}$ Oe cm² is the flux quantum. This means that if we have a superconducting cylinder placed in a magnetic field then decrease the temperature to a point below the superconducting transition temperature a flux will be trapped in the hole which will be an integral multiple of the flux quantum. If such an experiment were done in different starting values of the magnetic field, then the trapped flux will show quantum jumps with increasing applied field. In general, the current $J$ need not be zero.

$$\frac{m^*}{q\hbar}\oint \frac{\vec{J}}{|\psi|^2}\cdot d\vec{l} = \oint \left(\vec{\nabla}\varphi - \frac{q\vec{A}}{\hbar c}\right)\cdot d\vec{l} = 2\pi n - 2\pi\frac{\phi}{\phi_0}$$

or

$$\phi + \frac{m^*\phi_0}{|\psi|^2 qh}\oint \vec{J}\cdot d\vec{l} = n\phi_0$$

On simplification this gives

$$\phi + \frac{4\pi}{c}\lambda^2 \oint \vec{J}\cdot d\vec{l} = n\phi_0$$

Here, the LHS is called the fluxoid and the condition is referred to as the fluxoid quantisation. The above predictions have been verified experimentally by Deaver and Fairbank (Physical Review Letters 7, 43 (1961)) and Doll and Naebauer.

## Critical Field and Magnetisation of a Thin Film

Consider a thin superconducting film in a parallel field along the $z-$ direction $H_0$. The film is infinite in the $yz$ plane and extends from $x = -\frac{d}{2} to\, x = \frac{d}{2}$ or in reduced units, from $x = -\frac{d}{2\lambda} to\, x = \frac{d}{2\lambda}$. Here $A$ and $f$ will depend only on $x$. Also, $A$ is perpendicular to $x$. The first GL eqn in reduced units is then

$$\frac{d^2 f}{dx^2} = \kappa^2 f(A^2 - 1 + f^2)$$

Earlier, we have solved this (for a semi-infinite slab; infinite in the $yz$ plane and extending from 0 to $\infty$ along the $x$ direction) in the limit of small $\kappa$ by neglecting the RHS altogether to obtain $H = H_0 e^{-\frac{x}{\lambda}}$ etc. Now, our slab has both a lower and an upper bound in the $x-$ direction. So the solution will be a combination of rising and falling exponentials. The solution written in convenional units is given below

$$H = H_0 \frac{\cosh(f\frac{x}{\lambda})}{\cosh\frac{fd}{2\lambda}}$$

$$A = \frac{\lambda H_0}{f} \frac{\sinh(f \frac{x}{\lambda})}{\cosh \frac{fd}{2\lambda}}$$

For thin films ( $d \ll \lambda$) $A \simeq H_0 x$ . Again, neglecting the gradient of the order parameter, we will get (in conventional units) $f^2 = 1 - \frac{A^2}{2H_c^2 \lambda^2}$. Substituting $A \simeq H_0 x$ and then averaging over the width of the slab gives $< f^2 >= 1 - \frac{1}{24} \frac{H_0^2 d^2}{H_c^2 \lambda^2}$. Therefore, the order parameter will be zero when the applied field is $\sqrt{24} \frac{\lambda}{d} H_c$. As a result, for thin films, the actual critical field can be much larger than the thermodynamic critical field. We will now determine the magnetisation behaviour for thin and thick films. Once again, consider that $f$ is nearly a constant. The next level of approximation is to take $f = f + f_1$ where any $x$-dependence of the order parameter is in $f_1$. Substitute this back into the first GL equation (Eqn. 42). Now the differential equation is

$$\frac{d^2 f_1}{dx^2} = \kappa^2 f (A^2 - 1 + f^2)$$

For this inhomogeneous equation to have a solution, it is necessary that the RHS be orthogonal to the solution of homogeneous equation with correct boundary conditions; such a solution is $f_1 =$ constant. Hence, limited by the slab region, $\int (A^2 - 1 + f^2) dx = 0$. (In the zeroth approx. we neglect the RHS altogether while in the first approx. we keep the RHS but consider $f$ to be $x-$ independent and the boundary condition is that $\frac{df}{dx} = 0$, so we integrate once and set that equal to zero.) This leads to (just substitute for $A = \frac{H_0}{f} \frac{\sinh(fx)}{\cosh \frac{fd}{2}}$ and integrate between the given limits)

$$H_0^2 = \frac{2f^2(1-f^2)\cosh^2(\frac{fd}{2})}{\frac{\sinh fd}{fd} - 1}$$

We can then write the magnetisation of the film (in units of $H_c\sqrt{2}$) as

$$4\pi M = B - H_0 = \frac{1}{d}\int H dx - H_0 = -H_0[1 - \frac{\tanh \frac{fd}{2}}{\frac{fd}{2}}].$$

For $fd \gg 1, 4\pi M = -H_0$ which is the full Meissner effect. In the other extreme (thin film) $fd \ll 1, \tanh x = x - \frac{x^3}{3}$

$$4\pi M = -\frac{1}{12} H_0 d^2 f^2,$$

$$H_0^2 = \frac{12(1-f^2)}{d^2}$$

$$\therefore 4\pi M = -\frac{1}{12} H_0 d^2 (1 - \frac{H_0 d^2}{12})$$

In conventional units, this will be

$$4\pi M = -\frac{1}{12} H_0 (d/\lambda)^2 (1 - \frac{H_0 (d/\lambda)^2}{12})$$

## Nucleation of Superconductivity in Bulk Samples: Solution Via Linearised GL Equations

In the first GL equation, if we keep only the terms linear in $\psi$ we get,

$$(\frac{\vec{\nabla}}{i} - \frac{2\pi \vec{A}}{\phi_0})^2 \psi = -\frac{2m^* \alpha}{\hbar^2} \psi \equiv \frac{\psi}{\xi^2(T)}$$

Further, we consider the vector potential as that corresponding to the applied field. In other words, the screening effects are ignored. The above equation is like a Schroedinger equation for a charged particle in a magnetic field and hence can be solved by known techniques. Consider a superconductor in an applied field $H$ along the $z-$axis. A suitable vector potential which is commensurate with this is $A_y = Hx.$

$$[-\nabla^2 + \frac{4\pi i}{\phi_0} Hx \frac{\partial}{\partial y} + (\frac{2\pi H}{\phi_0})^2 x^2] = \frac{1}{\xi^2(T)} \psi$$

We subtitute in the above equation a solution of the form

$$\psi = e^{ik_y y} e^{ik_z z} f(x)$$

Simplifying we get

$$-f''(x) + (\frac{2\pi H}{\phi_0})^2 (x - x_0)^2 f = (\frac{1}{\xi^2} - k_z^2) f$$

In this equation, we have defined $x_0 = \frac{k_y \phi_0}{2\pi H}$. It can be seen that the above equation is similar to the Schroedinger equation for a particle in a harmonic oscillator potential with a suitably defined "force constant". The energy eigen values are given by

$$E_n = (n+\frac{1}{2})\hbar(\frac{2eH}{m^*c})$$

This should be equated to $(\frac{\hbar^2}{2m^*})(\frac{1}{\xi^2}-k_z^2)$. Simplifying and solving for $H$, we get

$$H = \frac{\phi_0}{2\pi(2n+1)}(\frac{1}{\xi^2}-k_z^2)$$

The highest value of the applied field consistent with this equation is obtained when $n=0$ and $k_z=0$. We get

$$H_{c2} = \frac{\phi_0}{2\pi\xi^2(T)}$$

$H_{c2}$ is known as the upper critical field.

## Magnetic Poperties of Classic Type-II Superconductors

As seen earlier, superconductors with $\kappa > \frac{1}{\sqrt{2}}$ have solutions of the GL equations with non-zero order parameter for filds less than $H_{c2}$. The Abrikosov solution has a regular array of vortices. Each unit cell of the flux line array has total flux equal to $\phi_0 = \frac{hc}{2e}$. We will now examine the solution to the GL equation at $H_{c1}$ where the first vortex enters the superconductor.

## Field and Order Parameter Variation Inside a Vortex

Consider a situation where the applied field is small such that the vortex density is small. No interaction between vortices is considered. As argued earlier, the term in a magnetic system, analogous to the $-PdV$ term of a hydrostatic system, is $HdM$. Therefore, when $H$ is the control parameter (held constant), the relevant free energy is the Gibbs free energy. At $H = H_{c1}$

$$G_s\,|_{\,no\,flux} = G_s\,|_{\,first\,vortex}$$

Note that $G = F - \frac{H}{4\pi}\int hd^3r$ where $h$ is the $B$-field at the location $r$ For the case of no flux in the superconductor, $G_s = F_s$. The LHS of equation (1) is then $F_s$. Imagine that the vortex has an extra free energy per unit length $\epsilon_1$. For a vortex of length $L$, the energy contribution is $\epsilon_1 L$. Therefore, RHS of equation (1) is $F_s - \frac{H_{c1}}{4\pi}\int hd^3r + \epsilon_1 L$ Remember that h is the local $B-$field. Hence $\int hd^3r$ (which is integrated over the volume of the vortex) is $\phi_0 L$ where $\phi_0$ is the flux quantum. The energy lowering by the $\int hd^3r$ term is matched by the increase due to the vortex energy $\epsilon_1 L$ such that the net change in $G$ is zero. This leads to

$$H_{c1} = \frac{4\pi\epsilon_1}{\phi_0}$$

Now consider the extreme Type-II limit $\kappa \gg 1, i.e., \lambda \gg \xi$. The core of the vortex can be thought to have a radius $\xi$. Since the B-field falls over a length $\lambda$ (starting from the centre of the vortex) which is much greater than $\xi$, over most of the vortex the order parameter is at its full value and the superconductor can be treated as a London superconductor.

Let $\psi = \psi_\infty f(r)e^{i\theta}$ represent the vortex wavefunction (axial symmetry is built into this). Now the vector potential

$$\vec{A} = A(r)\hat{\theta} \text{ and } A(r) = \frac{1}{r}\int_0^r rh(r)dr$$

In the London gauge $\vec{\nabla}\bullet\vec{A} = 0$. Here that is not the case since $A_\infty$ here is $\frac{\phi_0}{2\pi r}$ while in the London gauge $A_\infty = 0$. Also, near the center of the vortex $A(r) \backsimeq \frac{h(0)r}{2}$. This is because $A(r) = \frac{1}{r}\int_0^r rh(r)dr = \frac{1}{r}\frac{r^2}{2}h(0)$. Now substitute $\psi = \psi_\infty f(r)e^{i\theta}$ in the first GL equation given below

$$\alpha\psi + \beta|\psi|^2\psi + \frac{1}{2m^*}|(-i\hbar\vec{\nabla} - \frac{q\vec{A}}{c})^2\psi = 0$$

Writing the gradient in cylindrical coordinates and then simplifying one gets

$$f - f^3 - \xi^2[(\frac{1}{r} - \frac{2\pi A}{\phi_0})^2 f - \frac{1}{r}\frac{d}{dr}(r\frac{df}{dr})] = 0$$

Similarly, write the equation for the current density $\vec{J} = \frac{c}{4\pi}\vec{\nabla}\times\vec{h}$. This gives

$$J_\theta = -\frac{c}{4\pi}\frac{d}{dr}(\frac{1}{r}\frac{d}{dr}(\text{rA})) = \frac{q\hbar}{m^*}\psi_\infty^2 f^2(\frac{1}{r} - \frac{2\pi A}{\phi_0})$$

We need simultaneous solutions to the two GL equations. This requires numerical methods for a general case. Consider the limiting situation $r \to 0$ i.e., near the centre of the vortex. Here $A(r) = \frac{h(0)r}{2}$. Substitute this in the first GL equation obtained above. Further, assume a power law solution $f = cr^n$ where $n$ is positive. For small $r$, the leading term will be $-\xi^2 cr^{n-2}(1-n^2)$. this should go to zero for small $r$. Therefore $n = 1$ *and* $f$ varies linearly near the centre of the vortex. To determine higher order corrections, consider that $f$ has terms quadratic and cubic in $r$ in addition to the linear one. This leads to

$$f \approx cr\{1 - \frac{r^2}{8\xi^2}[1 + \frac{h(0)}{H_{c2}}]\}$$

where $c$ is a normalisation constant. This shows that the rise of $f(r)$ begins to saturate at $r \approx 2\xi$. The approximate general solution is $f \approx \tanh\frac{vr}{\xi}$ where $v$ is a constant of order 1.

## References


- Tinkham, M. (1996). Introduction to Superconductivity, Second Edition. New York, NY: McGraw-Hill. ISBN 0486435032
- Ginzburg VL (July 2004). "On superconductivity and superfluidity (what I have and have not managed to do), as well as on the 'physical minimum' at the beginning of the 21 st century". Chemphyschem. 5 (7): 930–945. PMID 15298379. doi:10.1002/cphc.200400182
- Annett, James (2004). Superconductivity, Superfluids and Condensates. New York: Oxford university press. p. 62. ISBN 978-0-19-850756-7
- David J. E. Callaway (1990). "On the remarkable structure of the superconducting intermediate state". Nuclear Physics B. 344 (3): 627–645. Bibcode:1990NuPhB.344..627C. doi:10.1016/0550-3213(90)90672-Z
- Abrikosov, A. A. (1957). The magnetic properties of superconducting alloys. Journal of Physics and Chemistry of Solids, 2(3), 199–208
- Tinkham, M. (1996). Introduction to Superconductivity, Second Edition. New York, NY: McGraw-Hill. ISBN 0486435032
- Rjabinin, J. N.; Shubnikow, L. W. (1935). "Magnetic Properties and Critical Currents of Supra-conducting Alloys". Nature. 135 (3415): 581. Bibcode:1935Natur.135..581R. doi:10.1038/135581a0
- Lev D. Landau; Evgeny M. Lifschitz (1984). Electrodynamics of Continuous Media. Course of Theoretical Physics. 8. Oxford: Butterworth-Heinemann. ISBN 0-7506-2634-8

4

# Cooper Pair and Microscopic Theory in Superconductivity

Cooper pairs are pairs of electrons that bind together at lower temperature. It is fundamental to the property of superconductivity. The topics discussed in the chapter are of great importance to broaden the existing knowledge on superconductivity.

## Cooper Pair

In condensed matter physics, a Cooper pair or BCS pair is a pair of electrons (or other fermions) bound together at low temperatures in a certain manner first described in 1956 by American physicist Leon Cooper. Cooper showed that an arbitrarily small attraction between electrons in a metal can cause a paired state of electrons to have a lower energy than the Fermi energy, which implies that the pair is bound. In conventional superconductors, this attraction is due to the electron–phonon interaction. The Cooper pair state is responsible for superconductivity, as described in the BCS theory developed by John Bardeen, Leon Cooper, and John Schrieffer for which they shared the 1972 Nobel Prize.

Although Cooper pairing is a quantum effect, the reason for the pairing can be seen from a simplified classical explanation. An electron in a metal normally behaves as a free particle. The electron is repelled from other electrons due to their negative charge, but it also attracts the positive ions that make up the rigid lattice of the metal. This attraction distorts the ion lattice, moving the ions slightly toward the electron, increasing the positive charge density of the lattice in the vicinity. This positive charge can attract other electrons. At long distances, this attraction between electrons due to the displaced ions can overcome the electrons' repulsion due to their negative charge, and cause them to pair up. The rigorous quantum mechanical explanation shows that the effect is due to electron–phonon interactions, with the phonon being the collective motion of the positively-charged lattice.

The energy of the pairing interaction is quite weak, of the order of $10^{-3}$ eV, and thermal energy can easily break the pairs. So only at low temperatures, in metal and other substrates, are a significant number of the electrons in Cooper pairs.

The electrons in a pair are not necessarily close together; because the interaction is long range, paired electrons may still be many hundreds of nanometers apart. This distance is usually greater than the average interelectron distance, so many Cooper pairs can occupy the same space. Electrons have spin-$\frac{1}{2}$, so they are fermions, but the total spin of a Cooper pair is integer (0 or 1) so it is a composite boson. This means the wave functions are symmetric under particle interchange. Therefore unlike electrons, multiple Cooper pairs are allowed to be in the same quantum state.

The BCS theory is also applicable to other fermion systems, such as helium-3. Indeed, Cooper pairing is responsible for the superfluidity of helium-3 at low temperatures. It has also been recently demonstrated that a Cooper pair can comprise two bosons. Here, the pairing is supported by entanglement in an optical lattice.

### Relationship to Superconductivity

The tendency for all the Cooper pairs in a body to "condense" into the same ground quantum state is responsible for the peculiar properties of superconductivity.

Cooper originally considered only the case of an isolated pair's formation in a metal. When one considers the more realistic state of many electronic pair formations, as is elucidated in the full BCS theory, one finds that the pairing opens a gap in the continuous spectrum of allowed energy states of the electrons, meaning that all excitations of the system must possess some minimum amount of energy. This *gap to excitations* leads to superconductivity, since small excitations such as scattering of electrons are forbidden. The gap appears due to many-body effects between electrons feeling the attraction.

R. H. Ogg was first to suggest that electrons might act as pairs coupled by lattice vibrations in the material. This was indicated by the isotope effect observed in superconductors. The isotope effect showed that materials with heavier ions (different nuclear isotopes) had lower superconducting transition temperatures. This can be explained by the theory of Cooper pairing: heavier ions are harder for the electrons to attract and move (how Cooper pairs are formed), which results in a smaller binding energy for the pairs.

The theory of Cooper pairs is quite general and does not depend on the specific electron-phonon interaction. Condensed matter theorists have proposed pairing mechanisms based on other attractive interactions such as electron–exciton interactions or electron–plasmon interactions. Currently, none of these alternate pairing interactions has been observed in any material.

It should be mentioned that Cooper pairing does not really involve individual electrons pairing up to form "quasi-bosons". The paired states are energetically favored, and electrons go in and out of those states preferentially. This is a fine distinction that John Bardeen makes:

*"The idea of paired electrons, though not fully accurate, captures the sense of it."*

The mathematical description of the second-order coherence involved here is given by Yang.

## Cooper-Pair Problem

The central idea in the formulation of a microscopic theory of superconductivity is derived from the work of Cooper which showed that there was a bound state for two electrons interacting through an arbitrarily weak attractive potential, just above the filled Fermi sea. In this chapter we derive the energy of the bound state of two electrons for a model attractive potential by solving the corresponding Schroedinger equation, taking the two electrons to be just above the filled Fermi sea. This problem is now known as the Cooper-pair problem.

## Schroedinger Equation for Two Electrons Interacting Via an Attractive Potential

Consider two electrons just above a filled Fermi sea at $T = 0^0 K$. Let $\vec{r_1}$ and $\vec{r_2}$ be the position vectors of the two electrons interacting via a potential V $(\vec{r_1} - \vec{r_2})$. Note that this is not the coulomb interaction. We want to know if the two electrons can form a bound pair. Writing down the Schroedinger equation,

$$\left[-\frac{\hbar^2}{2m}\vec{\nabla}_1^2 - \frac{\hbar^2}{2m}\vec{\nabla}_2^2 + V(\vec{r_1} - \vec{r_2})\right]\Psi(\vec{r_1}s_1, \vec{r_2}s_2)$$

The wave function of the two-electron system consists of a spatial part $\Psi(\vec{r_1}, \vec{r_2})$ and a spin part $\chi_s(s_1, s_2)$,

$$\Psi(\vec{r_1}s_1, \vec{r_2}s_2) = \Psi(\vec{r_1}, \vec{r_2})\chi_s(s_1, s_2)$$

The spin wave function for the two electrons can be a singlet state,

$$\chi_s(s_1, s_2) = \frac{1}{\sqrt{2}}(|\uparrow\downarrow\rangle - \langle\uparrow\downarrow|)$$

or a spin triplet,

$$\chi_s(s_1, s_2) = \begin{cases} |\uparrow\uparrow\rangle \\ \frac{1}{\sqrt{2}}(|\uparrow\downarrow\rangle + \langle\uparrow\downarrow|) \\ |\downarrow\downarrow\rangle \end{cases}$$

Most of the superconductors, to date, are found to be in the spin singlet state, while some may be in the spin triplet states. However, in the following, we assume that the spin wave function corresponds to the singlet state. Once we have chosen the spin part of the wave function to be an odd function of the two spins the symmetry of the spatial part is determined by the requirement of antisymmetry of the total wave function of the two, identical, electrons,

$$\Psi(\vec{r_1}s_1, \vec{r_2}s_2) = -\Psi(\vec{r_2}s_2, \vec{r_1}s_1)$$

In this case, it implies that the spatial part of the wave function is symmetric with respect to the interchange of the position coordinates of the two electrons.

Once we have chosen the spatial and spin parts of the wave function, we make a transformation to the center of mass coordinate $\vec{R} = \frac{(\vec{r_1} + \vec{r_2})}{2}$ and the relative coordinate $\vec{r} = \vec{r_1} - \vec{r_2}$ so that

$$\Psi(\vec{r_1}, \vec{r_2}) \rightarrow \Psi(\vec{r}, \vec{R})$$

where for brevity we have kept the same symbol for the functional dependence in the two cases and the reduced mass $\mu$ is given by

$$\frac{1}{\mu} = \frac{1}{m} + \frac{1}{m} = \frac{2}{m}$$

$$or\ \mu = \frac{m}{2};\ or\ m = 2\mu.$$

The Schroedinger's equation can be rewritten in terms of $\vec{R}$ and $\vec{r}$ as

$$\left[ -\frac{\hbar^2}{2(2m)} \vec{\nabla}_{\vec{R}}^2 - \frac{\hbar^2}{2\left(\frac{m}{2}\right)} \vec{\nabla}_{\vec{r}}^2 + V(\vec{r}) \right] \Psi(\vec{r}, \vec{R}) = E\Psi(\vec{r}, \vec{R})$$

Using the separation of variables, the wave function of the two electrons can be separated into two parts as we will next.

Assuming that the wavefunction of the two elecrons can be written as

$$\Psi(\vec{r}, \vec{R}) = \psi(\vec{r})\phi(\vec{R})$$

we get after substituting it back in the Schroedinger equation,

$$\left[ -\frac{\hbar^2}{4m} \vec{\nabla}_{\vec{R}}^2 - \frac{\hbar^2}{m} \vec{\nabla}_{\vec{r}}^2 + V(\vec{r}) \right] \psi(\vec{r})\phi(\vec{R}) = E\psi(\vec{r})\phi(\vec{R})$$

Opening up the square bracket and dropping the arguments of the function for brevity, we have

$$-\frac{\hbar^2}{4m} \psi \vec{\nabla}_{\vec{R}}^2 \phi - \frac{\hbar^2}{m} \phi \vec{\nabla}_{\vec{r}}^2 \psi + V(\vec{r})\psi\phi = E\psi\phi$$

Dividing both sides by $\psi\phi$, we get

$$\left\{ -\frac{\hbar^2}{4m} \frac{1}{\phi} \vec{\nabla}_{\vec{R}}^2 \phi \right\} + \left\{ -\frac{\hbar^2}{m} \frac{1}{\psi} \vec{\nabla}_{\vec{r}}^2 \psi + V(\vec{r}) \right\} = E$$

The first term is independent of the relative coordinate $\vec{r}$, while the term in the second bracket is independent of the center of mass coordinate $\vec{R}$. The sum of the two terms is equal to a constant $E$.

Therefore, the two terms of this equation must be separately equal to some other constant, say $C$. The first term can be written as

$$-\frac{\hbar^2}{4m} \frac{1}{\phi} \vec{\nabla}_{\vec{R}}^2 \phi = C$$

or,

$$\vec{\nabla}^2_{\vec{R}}\phi + \frac{4mC}{\hbar^2}\phi = 0$$

or,

$$\vec{\nabla}^2_{\vec{R}}\phi + Q^2\phi = 0$$

where we have defined $Q^2 \equiv \frac{4mC}{\hbar^2}$. The solutions of this differential equation are given by plane waves, $\phi(\vec{R})$,

$$\phi(\vec{R}) = e^{i\vec{Q}\cdot\vec{R}}$$

In order to check that indeed the plane waves solve the above differential equation, we try in the Cartesian coordinate system,

$$or\ ,\left\{\frac{\partial^2}{\partial x^2}+\frac{\partial^2}{\partial y^2}+\frac{\partial^2}{\partial z^2}\right\}e^{i\vec{Q}.\vec{R}} = \left\{\frac{\partial^2}{\partial x^2}+\frac{\partial^2}{\partial y^2}+\frac{\partial^2}{\partial z^2}\right\}e^{i(Q_xX+Q_yY+Q_zZ)} = -Q^2e^{i(Q_xX+Q_yY+Q_zZ)}$$

$$or\ ,\left\{\frac{\partial^2}{\partial x^2}+\frac{\partial^2}{\partial y^2}+\frac{\partial^2}{\partial z^2}\right\}e^{i\vec{Q}.\vec{R}}$$

$$= \left\{\frac{\partial^2}{\partial x^2}+\frac{\partial^2}{\partial y^2}+\frac{\partial^2}{\partial z^2}\right\}e^{i(Q_xX+Q_yY+Q_zZ)}$$

$$= -Q^2e^{i(Q_xX+Q_yY+Q_zZ)}$$

where we have used, $e^{i\vec{Q}\cdot\vec{R}} = e^{i(Q_xX+Q_yY+Q_zZ)}$.

Now we can write down the differential equation involving the relative coordinate $\vec{r}$, by using

$$C = \frac{\hbar^2Q^2}{4m}$$

We get,

$$\frac{\hbar^2Q^2}{4m} + \left\{-\frac{\hbar^2}{m}\frac{1}{\psi}\vec{\nabla}^2_r\psi + V(\vec{r})\right\} = E$$

Taking the first term to the right, we have

$$\left\{-\frac{\hbar^2}{m}\vec{\nabla}^2_r + V(\vec{r})\right\}\psi = \left\{E - \frac{\hbar^2Q^2}{4m}\right\}\psi$$

In order to simplify this equation further, we note that the two electrons with momentum $\vec{P}_1$ and $\vec{P}_2$ lead to a center-of-mass momentum $\vec{Q} = \vec{P}_1 + \vec{P}_2$.

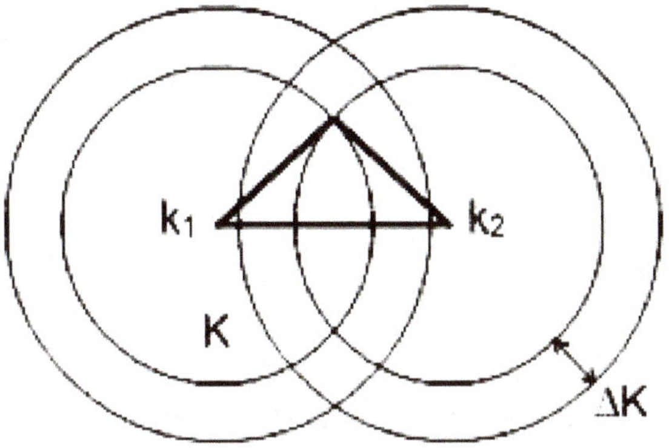


Two electrons with wave vectors $\vec{k}_1$ and $\vec{k}_2$, suffer collisions such that $\vec{k}_1 + \vec{k}_2 = \vec{k}'_1 + \vec{k}'_2 = \vec{k}$, i.e., the net momentum remains unchanged. The two spherical shells shown in the figure are in the reciprocal space with radii corresponding to $\in_F - \hbar\omega_D$ and $\in_F + \hbar\omega_D$ respectively. The number of electron pairs with $\vec{k}_1 + \vec{k}_2 = \vec{k}$ is proportional to the $\vec{k}$-space volume corresponding to the overlap of the two shells (shown shaded in the figure). Note that the overlap is maximum for $\vec{k} = 0$.

## Cooper-pair with Zero Center-of-mass Momentum

For zero center-of-mass momentum, the Schroedinger equation for the relative motion of the two electrons become

$$-\frac{\hbar^2}{m}\vec{\nabla}_r^2\psi + V(\vec{r})\psi = E\psi$$

Since the two electrons are above the filled Fermi Sea, the spatial part of the wave function can be expanded in terms of plane waves,

$$\psi(\vec{r}) = \frac{1}{\sqrt{V}}\sum_{k>k_F} g_{\vec{k}} e^{i\vec{k}\cdot\vec{r}}$$

As we are considering symmetric spatial part, i.e.,

$$\psi(\vec{r}_1 - \vec{r}_2) = +\psi(\vec{r}_2 - \vec{r}_1)$$

which requires that

$$\frac{1}{\sqrt{V}}\sum_{k>k_F} g_{\vec{k}} e^{i\vec{k}.(\vec{r}_1-\vec{r}_2)} = \frac{1}{\sqrt{V}}\sum_{k>k_F} g_{\vec{k}} e^{i\vec{k}\cdot(\vec{r}_2-\vec{r}_1)}$$

For each $\vec{k}$, we should have,

$$g_{\vec{k}} e^{i\vec{k}\cdot(\vec{r}_1-\vec{r}_2)} = g_{\vec{k}} e^{-i\vec{k}\cdot(\vec{r}_1-\vec{r}_2)}$$

The above equation holds if $g_{\vec{k}} = g_{-\vec{k}}$.

Substituting the above expansions of $\psi(\vec{r})$ into the Schroedinger equation,

$$-\frac{\hbar^2}{m}\vec{\nabla}^2\frac{1}{\sqrt{V}}\sum_{\vec{k}'} g_{\vec{k}'}e^{i\vec{k}'\cdot\vec{r}} + \frac{1}{\sqrt{V}}\sum_{\vec{k}'} V(\vec{\mathrm{r}})e^{i\vec{k}'\cdot\vec{r}}\, g_{\vec{k}} = \frac{1}{\sqrt{V}}E\sum_{\vec{k}'} g_{\vec{k}'}e^{i\vec{k}'\cdot\vec{r}}$$

Rearranging the two sides, we get

$$\frac{1}{\sqrt{V}}\sum_{\vec{k}'} g_{\vec{k}'}\left\{-\frac{\hbar^2 k'^2}{m}+E\right\}e^{i\vec{k}',\cdot\vec{r}} = \frac{1}{\sqrt{V}}\sum_{\vec{k}'} g_{\vec{k}'}V(\vec{r})e^{i\vec{k}'\cdot\vec{r}}$$

Multiplying from left by $e^{-i\vec{k}\cdot\vec{r}}$ and integrating over the volume, we get

$$\frac{1}{\sqrt{V}}\sum_{\vec{k}'} g_{\vec{k}'}\left\{E-2\varepsilon_{\vec{k}'}\right\}\int e^{-i(\vec{k}-\vec{k}')\cdot\vec{r}}d^3r = \frac{1}{\sqrt{V}}\sum_{\vec{k}'} g_{\vec{k}'}\frac{1}{\sqrt{V}}\int e^{-i\vec{k}\vec{r}}V(\vec{r})e^{-i\vec{k}'\cdot\vec{r}}d^3r$$

Defining Fourier coefficient of the potential $V(\vec{r})$ as

$$V_{\vec{k}\vec{k}'} = \frac{1}{V}\int e^{-i(\vec{k}-\vec{k}')\cdot\cdot r}V(\vec{r})d^3\vec{r}$$

and using

$$\frac{1}{V}\int e^{-i(\vec{k}-\vec{k}')\cdot\vec{r}}d^3\vec{r} = \delta_{\vec{k}\vec{k}'}$$

the Schroedinger equation reduces to

$$\sum_{\vec{k}'} g_{\vec{k}'}\left\{E-2\varepsilon_{\vec{k}'}\right\}\delta_{\vec{k}\vec{k}'} = \sum_{\vec{k}'} g_{\vec{k}'}V_{\vec{k}\vec{k}'}$$

Carrying out the summation over $\vec{k}'$ in the left-hand side of the above equation and rewriting, we have

$$\sum_{\vec{k}'} g_{\vec{k}'}V_{\vec{k}\vec{k}'} = g_{\vec{k}}(E-2\varepsilon_{\vec{k}})$$

In order to proceed further, we have to specify the matrix elements of the potential. A very simple approximation for the matrix elements of the potential is to assume it to be attractive in a narrow range of energy around the Fermi energy such that

$$V_{\vec{k}\vec{k}'} = \begin{cases} -V_0, & \varepsilon_F < \varepsilon_{\vec{k}} < \varepsilon_F + \hbar\omega_D \; and \; \varepsilon_F < \varepsilon_{\vec{k}'} < \varepsilon_F + \hbar\omega_D \\ 0, & otherwise \end{cases}$$

For $\varepsilon_{\vec{k}}$ and $\varepsilon_{\vec{k}'}$ specified as above, we get

$$-V_0\sum_{\vec{k}'} g_{\vec{k}'} = g_{\vec{k}}(E-2\varepsilon_{\vec{k}})$$

or,

$$g_{\vec{k}} = \frac{-V_o}{E - 2\varepsilon_{\vec{k}}} \sum_{\vec{k}'} g_{\vec{k}'}$$

This can be further simplified by summing over $\vec{k}$,

$$\sum_{\vec{k}} g_{\vec{k}} = \sum_{\vec{k}} \frac{-V_o}{E - 2\varepsilon_{\vec{k}}} \sum_{\vec{k}'} g_{\vec{k}'}$$

Since $\sum_{\vec{k}} g_{\vec{k}} = \sum_{\vec{k}'} g_{\vec{k}'}$ we are left with

$$-\sum_{\vec{k}} \frac{V_o}{E - 2\varepsilon_{\vec{k}}} = 1$$

After rearranging,

$$V_0 \sum_{\vec{k}} \frac{1}{2\varepsilon_{\vec{k}} - E} = 1$$

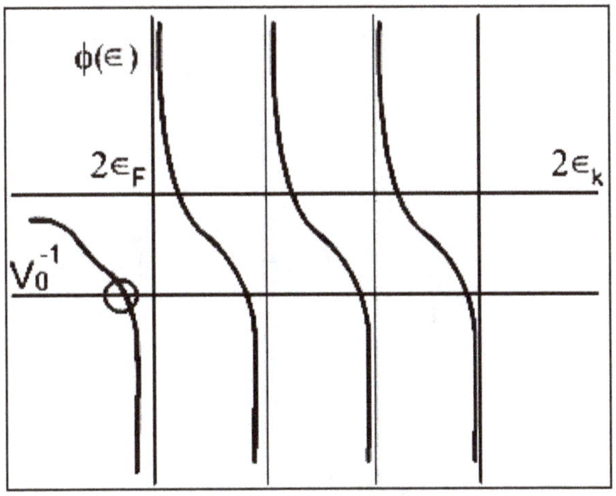


The function $\phi(E) = \sum_k \frac{1}{2\varepsilon_k - E}$ goes from $-\infty$ to $+\infty$ as it crosses the eigenvalues $2\varepsilon_k$ above as shown in the figure. As $E$ passes from origin to negative values, $\phi(E)$ increases from $-\infty$ to $0$. The intersection points of $\frac{1}{\lambda}$ with $\phi(E)$ correspond to the solution of the equation. Note that for an attractive interaction implying a negative $\lambda$, no matter how small, then is always a bound state below the continum.

## Cooper-pair with Bound States

To determine the energy eigenvalues of the two electron system analytically from the above equation, we change the summation with respect to wave vector $\vec{k}$ to integration with respect to energy $\vec{E}$ using the density of states $N(\varepsilon)$,

$$\frac{1}{N}\sum_{\vec{k}}=\frac{V/N}{(2\pi)^3}\int d^3\vec{k}\rightarrow\int N(\varepsilon)d\varepsilon$$

where $N(\varepsilon)$ is the density of states for one spin per unit cell (We have $N$ such unit cells), we get

$$1=V_0\int_{\varepsilon_F}^{\varepsilon+\hbar\omega_D}\frac{N(\varepsilon)}{2\varepsilon-E}d\varepsilon$$

where $\omega_D$ is the Debye frequency. Since the range of integration is very small, it is a good approximation to assume that the density of states is constant over this range, and take it to be equal to the density of states at the Fermi energy,

$$1=V_0N(\varepsilon_F)\frac{1}{2}\int_{\varepsilon_F}^{\varepsilon+\hbar\omega_D}\frac{d(2\varepsilon-E)}{2\varepsilon-E}$$

$$=\frac{V_0N(\varepsilon_F)}{2}ln\left|2\varepsilon-E\right|_{\varepsilon_F}^{\varepsilon_F+\hbar\omega_D}$$

$$=\frac{V_0N(\varepsilon_F)}{2}ln\left[\frac{2(\varepsilon_F+\hbar\omega_D)-E}{2\varepsilon_F-E}\right]$$

or,

$$\frac{2}{V_0N(\varepsilon_F)}=ln[1+\frac{2\hbar\omega_D}{2\varepsilon_F-E}]$$

or,

$$\frac{2}{V_0N(\varepsilon_F)}=ln\left[\frac{2\varepsilon_F-E+\hbar\omega_D}{2\varepsilon_F-E}\right]$$

Using the property of logarithms, we can write,

or,

$$-\frac{2}{V_0N(\varepsilon_F)}=ln\left[\frac{2\varepsilon_F-E}{2\varepsilon_F-E+\hbar\omega_D}\right]$$

or,

$$e^{-\frac{2}{V_0N(\varepsilon_F)}}=\frac{2\varepsilon_F-E}{2\varepsilon_F-E+\hbar\omega_D}$$

or,

$$e^{-\frac{2}{V_0N(\varepsilon_F)}}(2\varepsilon_F - E + \hbar\omega_D) = 2\varepsilon_F - E$$

In order to solve for $E$, Let $e^{-\frac{2}{V_0N(\varepsilon_F)}} \equiv X$, so that

$$(2\varepsilon_F + \hbar\omega_D)X - 2\varepsilon_F = (X-1)E$$

or,

$$E = \frac{(2\varepsilon_F + \hbar\omega_D)e^{-\frac{2}{V_0N(\varepsilon_F)}} - 2\varepsilon_F}{e^{-\frac{2}{V_0N(\varepsilon_F)}}}$$

or,

$$E = 2\varepsilon_F - \frac{2\hbar\omega_D e^{-\frac{2}{V_0N(\varepsilon_F)}}}{1 - e^{-\frac{2}{V_0N(\varepsilon_F)}}}$$

For $2 \gg V_0N(\varepsilon_F)$, the exponential in the denominator can be expanded and we are left with

$$E = 2\varepsilon_F - 2\hbar\omega_D e^{-\frac{2}{V_0N(\varepsilon_F)}}$$

which shows that the two-electron system is unstable with respect to pair formation as long as $V_0$ is positive making the interaction between the two electrons attractive. The resulting energy $E$ of the two electron system, which had its starting energy equal to $2\epsilon_F$, is reduced by $2\hbar\omega_D e^{-\frac{2}{V_0N(\varepsilon_F)}}$.

Thus, we see that if $V_0 > 0$ such that there is an attractive interaction between the two electrons, then there is always a solution for $E < 2\epsilon_F$. Such a solution leads to a bound state for the two electrons leading to Cooper pair formation.

## Spatial Extent of a Cooper-pair Wave Function

To see the spatial extent of the Cooper-pair wave function, we calculate its mean square radius ,$\overline{r^2}$, as follows.

$$\overline{r^2} = \frac{\int |\psi(\vec{r})|^2 r^2 d^3\vec{r}}{\int |\psi(\vec{r})|^2 d^3\vec{r}}$$

Using

$$\psi(\vec{r}) = \frac{1}{\sqrt{V}} \sum_{\vec{k}} g_{\vec{k}} e^{i\vec{k}\cdot\vec{r}}$$

we have

$$|\psi(\vec{r})|^2 = \frac{1}{V}\sum_{\vec{k}\vec{k}'} g_{\vec{k}} g_{\vec{k}'}^{*} e^{i\vec{k}-\vec{k}'\cdot\vec{r}}$$

Integrating over the volume yields

$$\int |\psi(\vec{r})|^2 d^3\vec{r} = \sum_{\vec{k}\vec{k}'} g_{\vec{k}} g_{\vec{k}'}^{*} \delta_{\vec{k}\vec{k}'}$$

Summing over $k'$ gives the denominator of Equation to be

$$\int |\psi(\vec{r})|^2 d^3\vec{r} = \sum_{\vec{k}} | g_{\vec{k}} |^2$$

For evaluating the numerator $\int |\psi(\vec{r})|^2 r^2 d^3\vec{r}$, we note that

$$-i\vec{\nabla}_{\vec{k}} e^{i\vec{k}\cdot\vec{r}} = r e^{i\vec{k}\cdot\vec{r}}$$

from which we get,

$$\int |\psi(\vec{r})|^2 r^2 d^3\vec{r} = \sum_{\vec{k}} | \vec{\nabla}_{\vec{k}} g_{\vec{k}} |^2$$

So that the mean square radius of a Cooper pair wavefunction is

$$\overline{r^2} = \frac{\sum_{\vec{k}} | \vec{\nabla}_{\vec{k}} g_{\vec{k}} |^2}{\sum_{\vec{k}} | g_{\vec{k}} |^2}$$

We have seen earlier that

$$g_{\vec{k}} \propto \frac{1}{2\varepsilon_{\vec{k}} - E}$$

and since

$$\frac{\partial}{\partial k} = \frac{\partial}{\partial \varepsilon_{\vec{k}}} \frac{\partial \varepsilon_{\vec{k}}}{\partial k} = \hbar v_F \frac{\partial}{\partial \varepsilon_{\vec{k}}}$$

where $v_F$ is the Fermi velocity.

Changing over to energy integration from summation in Equation, we have

$$\overline{r^2} = \frac{(\hbar v_F)^2 \int \left[ \frac{\partial}{\partial \varepsilon} \frac{1}{2\varepsilon - E} \right]^2 d\varepsilon}{\int \left[ \frac{1}{2\varepsilon - E} \right]^2 d\varepsilon}$$

$$= \frac{-\frac{2}{3}(\hbar v_F)^2 \left(\frac{1}{2\varepsilon - E}\right)^3 \Big|_0^\infty}{-\frac{1}{2}\frac{1}{2\varepsilon - E}\Big|_0^\infty}$$

Or,

$$\overline{r^2} = \frac{4}{3}\frac{(\hbar v_F)^2}{\varepsilon^2}$$

To get a feel for the spread of the Cooper-pair wave function, let us assume that $\varepsilon \sim k_B T_{C,}\, T_C \sim 10K$ and $v_F = 10^8$ cm/s, we get

$$\left(\overline{r^2}\right)^{1/2} \sim 10^{-4} cm$$

Or,

$$\left(\overline{r^2}\right)^{1/2} \sim 10^4 \mathring{A}$$

which should be compared with a typical volume occupied by an electron which is $\sim \left(2\mathring{A}\right)^3$. Thus, within the volume occupied by a pair of electrons there are around $10^{11}$ electrons in the same volume. The many-body wave function of the Bardeen-Cooper-Schrieffer theory must take this into account.

## Cooper-Pair Problem using Second Quantization Formalism

Consider the Cooper pair problem using the second quantization formalism. The wavefunction of the two electrons interacting above the Fermi sea of non-interacting electrons can be written as

$$|\ _{12}\rangle = \sum_{\vec{k}_1\vec{k}_2\sigma_1\sigma_2} g_{\vec{k}_1\vec{k}_2\sigma_1\sigma_2} C^+_{\vec{k}_1\sigma_1} C^+_{\vec{k}_2\sigma_2} |G>$$

where $|G>$ indicates the Fermi sea and $\sigma_1$ *and* $\sigma_2$ are the spin variables. The sum over $\vec{k}_1$ *and* $\vec{k}_2$ is carried out within the constraint $\vec{K} = \vec{k}_1 + \vec{k}_2$. As we have seen earlier, to maximize the effect of interaction we should take $\vec{K} = 0$. In the following, we take a spin-singlet state, knowing that it has lower energy than the triplet state, we can write the wavefunction for the electron pair as

$$|\psi_{12}\rangle = \sum_{\vec{k}} g\left(\vec{k}\right) C^+_{\vec{k}\sigma} C^+_{-\vec{k}-\sigma} |G>$$

To see if there is a bound state for the electron pair, we calculate the expectation value of energy using the Hamiltonian

$$H = \sum_{\vec{k}\sigma} \epsilon_{\vec{k}} C^{+}_{\vec{k}\sigma} c_{\vec{k}\sigma} - \frac{V}{2} \sum_{\vec{k}\vec{q}\sigma} C^{+}_{\vec{k}+\vec{q}\,\sigma}$$

where $V \neq 0$, and $\left|\epsilon_{\vec{k}+\vec{q}} - \epsilon_{\vec{k}}\right| \leq \hbar\omega_{\vec{q}}$ with $\omega_{\vec{q}}$ being a characteristic frequency of the phonon spectrum, say $\omega_D$, the Debye frequency. Now, the expectation value of energy can be evaluated by first considering the kinetic energy term,

$$<\psi_{12}\left|\epsilon_{\vec{k}} C^{+}_{\vec{k},\sigma} C_{\vec{k},\sigma}\right|\psi_{12}>=<G\left|\sum_{\vec{l}} g^{*}_{\vec{l}} C_{\vec{l},\sigma} C_{-\vec{l},-\sigma} \sum_{\vec{k}\sigma} \epsilon_{\vec{k}} C^{+}_{\vec{k}\sigma} C_{\vec{k}\sigma} \sum_{\vec{m}} g_{\vec{m}} C^{+}_{\vec{m}\sigma} C_{-\vec{m},-\sigma}\right|G>$$

Only $\vec{l} = \vec{m} = \vec{k}$ term survives and $\sigma$ is summed up for two values, leading to

$$= 2\sum_{\vec{k}} \epsilon_{\vec{k}} \left|g_{\vec{k}}\right|^{2}$$

Similarly for the interaction part, we get

$$-\frac{V}{2} <G\left|\sum_{\vec{l}} g^{*}_{\vec{l}} C_{\vec{l}\sigma} C_{-\vec{l},-\sigma} \sum_{\vec{k}\vec{q}\sigma} C^{+}_{\vec{k}+\vec{q}\sigma} C^{+}_{-\vec{k}-\vec{q},-\sigma} C_{-\vec{k},-\sigma} C_{\vec{k}\sigma} \sum_{\vec{m}} g_{\vec{m}} C^{+}_{\vec{m},\sigma} C^{+}_{-\vec{m},-\sigma}\right.$$

If $C_{\vec{l}}$, acting on the left, creates an electron with momentum $\vec{k} + \vec{q}$ and $C^{+}_{\vec{m}}$, acting on the right, creates an electron with momentum $\vec{k}$ then for up and down spins the interaction term simplifies to

$$-V\sum_{\vec{k}\vec{q}} g^{*}_{\vec{k}+\vec{q}} g_{\vec{k}}$$

Adding the kinetic energy and the interaction terms, the expectation value of energy is given by

$$E = 2\sum_{\vec{k}} \epsilon_{\vec{k}} \left|g_{\vec{k}}\right|^{2} - V\sum_{\vec{k}\vec{q}} g^{*}_{\vec{k}+\vec{q}} g_{\vec{k}}$$

In order to carry out the variational minimization of energy with the constraint

$$\sum_{\vec{k}} \left|g_{\vec{k}}\right|^{2} = 1$$

we use the Lagrange multiplier $\lambda$ to incorporate the above constraint and write the energy as

$$E' = \mathrm{E} - \lambda \sum_{\vec{k}} \left|g_{\vec{k}}\right|^{2}$$

Minimizing with respect to $g^{*}_{\vec{k}}$,

$$\frac{\partial E'}{\partial g^{*}_{\vec{k}}} = \frac{\partial}{\partial g^{*}_{\vec{k}}} \left[ 2\sum_{\vec{k}} \epsilon_{\vec{k}} \left|g_{\vec{k}}\right|^{2} - V\sum_{\vec{k}\vec{q}} g^{*}_{\vec{k}+\vec{q}} g_{\vec{k}} - \lambda \sum_{\vec{k}} \left|g_{\vec{k}}\right|^{2} \right]$$

$$= 2\epsilon_{\vec{k}} g_k - V\sum_{\vec{q}} g_{\vec{k}-\vec{q}} - \lambda g_{\vec{k}}$$

$$= 0$$

The second term on the right in the middle equation arises because $\vec{k} + \vec{q} = \vec{k}'$, only then it is non zero. Replacing $\vec{q}$ by $\vec{k}'$, we have

$$\left(2\epsilon_{\vec{k}} - \lambda\right) g_{\vec{k}} = V\sum_{\vec{k}} g_{\vec{k}}$$

Let $\sum_{\vec{k}} g_{\vec{k}} = C.$ Note that the summation is such that $\vec{k}$ corresponds to energies between $\varepsilon_F$ and $\varepsilon_F + \hbar\omega_{\vec{q}}.$ Now

$$\left(2\epsilon_{\vec{k}} - \lambda\right) g_{\vec{k}} = VC$$

or,

$$g_{\vec{k}} = \frac{VC}{\left(2\epsilon_{\vec{k}} - \lambda\right)}$$

### Microscopic Theory

A microscopic theory is one that contains an explanation at the atomic or subatomic level in contrast to a higher level or classical macroscopic or *phenomenological theory*. e.g. in superconductivity BCS theory is a microscopic theory.

## BCS Theory

BCS theory or Bardeen–Cooper–Schrieffer theory (named after John Bardeen, Leon Cooper, and John Robert Schrieffer) is the first microscopic theory of superconductivity since Heike Kamerlingh Onnes's 1911 discovery. The theory describes superconductivity as a microscopic effect caused by a condensation of Cooper pairs into a boson-like state. The theory is also used in nuclear physics to describe the pairing interaction between nucleons in an atomic nucleus.

It was proposed by Bardeen, Cooper, and Schrieffer in 1957; they received the Nobel Prize in Physics for this theory in 1972.

### History

Rapid progress in understanding superconductivity gained momentum in the mid-1950s. It began with the 1948 paper, "On the Problem of the Molecular Theory of Superconductivity", where Fritz London proposed that the phenomenological London equations may be consequences of the coherence of a quantum state. In 1953, Brian Pippard, motivated by penetration experiments, proposed that this would modify the London equations via a new scale parameter called the coherence length. John Bardeen then argued in the 1955 paper, "Theory of the Meissner Effect in Superconductors", that such a modification naturally occurs in a theory with an energy gap. The key ingredient was Leon Neil Cooper's calculation of the bound states of electrons subject to an attractive force in his 1956 paper, "Bound Electron Pairs in a Degenerate Fermi Gas".

In 1957 Bardeen and Cooper assembled these ingredients and constructed such a theory, the BCS theory, with Robert Schrieffer. The theory was first published in April 1957 in the letter, "Microscopic theory of superconductivity". The demonstration that the phase transition is second order, that it reproduces the Meissner effect and the calculations of specific heats and penetration depths appeared in the December 1957 article, "Theory of superconductivity". They received the Nobel Prize in Physics in 1972 for this theory.

In 1986, high-temperature superconductivity was discovered in some materials at temperatures up to about 130 K, considerably above the previous limit of about 30 K. It is believed that BCS theory alone cannot explain this phenomenon and that other effects are in play. These effects are still not yet fully understood; it is possible that they even control superconductivity at low temperatures for some materials.

## Overview

At sufficiently low temperatures, electrons near the Fermi surface become unstable against the formation of Cooper pairs. Cooper showed such binding will occur in the presence of an attractive potential, no matter how weak. In conventional superconductors, an attraction is generally attributed to an electron-lattice interaction. The BCS theory, however, requires only that the potential be attractive, regardless of its origin. In the BCS framework, superconductivity is a macroscopic effect which results from the condensation of Cooper pairs. These have some bosonic properties, and bosons, at sufficiently low temperature, can form a large Bose–Einstein condensate. Superconductivity was simultaneously explained by Nikolay Bogolyubov, by means of the Bogoliubov transformations.

In many superconductors, the attractive interaction between electrons (necessary for pairing) is brought about indirectly by the interaction between the electrons and the vibrating crystal lattice (the phonons). Roughly speaking the picture is the following:

An electron moving through a conductor will attract nearby positive charges in the lattice. This deformation of the lattice causes another electron, with opposite spin, to move into the region of higher positive charge density. The two electrons then become correlated. Because there are a lot of such electron pairs in a superconductor, these pairs overlap very strongly and form a highly collective condensate. In this "condensed" state, the breaking of one pair will change the energy of the entire condensate - not just a single electron, or a single pair. Thus, the energy required to break any single pair is related to the energy required to break *all* of the pairs (or more than just two electrons). Because the pairing increases this energy barrier, kicks from oscillating atoms in the conductor (which are small at sufficiently low temperatures) are not enough to affect the condensate as a whole, or any individual "member pair" within the condensate. Thus the electrons stay paired together and resist all kicks, and the electron flow as a whole (the current through the superconductor) will not experience resistance. Thus, the collective behavior of the condensate is a crucial ingredient necessary for superconductivity.

## Details

BCS theory starts from the assumption that there is some attraction between electrons, which can overcome the Coulomb repulsion. In most materials (in low temperature superconductors), this attraction is brought about indirectly by the coupling of electrons to the crystal lattice (as explained above). However, the results of BCS theory do *not* depend on the origin of the attractive interaction. For instance, Cooper pairs have been observed in ultracold gases of fermions where a homogeneous magnetic field has been tuned to their Feshbach resonance. The original results of BCS (discussed below) described an s-wave superconducting state, which is the rule among low-temperature superconductors but is not realized in many unconventional superconductors such as the d-wave high-temperature superconductors.

Extensions of BCS theory exist to describe these other cases, although they are insufficient to completely describe the observed features of high-temperature superconductivity.

BCS is able to give an approximation for the quantum-mechanical many-body state of the system of (attractively interacting) electrons inside the metal. This state is now known as the BCS state. In the normal state of a metal, electrons move independently, whereas in the BCS state, they are bound into Cooper pairs by the attractive interaction. The BCS formalism is based on the reduced potential for the electrons' attraction. Within this potential, a variational ansatz for the wave function is proposed. This ansatz was later shown to be exact in the dense limit of pairs. Note that the continuous crossover between the dilute and dense regimes of attracting pairs of fermions is still an open problem, which now attracts a lot of attention within the field of ultracold gases.

## Underlying Evidence

The hyperphysics website pages at Georgia State University summarize some key background to BCS theory as follows:

- Evidence of a band gap at the Fermi level (described as "a key piece in the puzzle")

  the existence of a critical temperature and critical magnetic field implied a band gap, and suggested a phase transition, but single electrons are forbidden from condensing to the same energy level by the Pauli exclusion principle. The site comments that "a drastic change in conductivity demanded a drastic change in electron behavior". Conceivably, pairs of electrons might perhaps act like bosons instead, which are bound by different condensate rules and do not have the same limitation.

- Isotope effect on the critical temperature, suggesting lattice interactions

  The Debye frequency of phonons in a lattice is proportional to the inverse of the square root of the mass of lattice ions. It was shown that the superconducting transition temperature of mercury indeed showed the same dependence, by substituting natural mercury $^{202}Hg$ with a different isotope $^{198}Hg$.

- An exponential rise in heat capacity near the critical temperature for some superconductors

  An exponential increase in heat capacity near the critical temperature also suggests an energy bandgap for the superconducting material. As superconducting vanadium is warmed toward its critical temperature, its heat capacity increases massively in a very few degrees; this suggests an energy gap being bridged by thermal energy.

- The lessening of the measured energy gap towards the critical temperature

  this suggests a type of situation where some kind of binding energy exists but it is gradually weakened as the critical temperature is approached. A binding energy suggests two or more particles or other entities that are bound together in the superconducting state. This helped to support the idea of bound particles - specifically electron pairs - and together with the above helped to paint a general picture of paired electrons and their lattice interactions.

## Implications

BCS derived several important theoretical predictions that are independent of the details of the interaction, since the quantitative predictions mentioned below hold for any sufficiently weak attraction between the electrons and this last condition is fulfilled for many low temperature superconductors - the so-called weak-coupling case. These have been confirmed in numerous experiments:

- The electrons are bound into Cooper pairs, and these pairs are correlated due to the Pauli exclusion principle for the electrons, from which they are constructed. Therefore, in order to break a pair, one has to change energies of all other pairs. This means there is an energy gap for single-particle excitation, unlike in the normal metal (where the state of an electron can be changed by adding an arbitrarily small amount of energy). This energy gap is highest at low temperatures but vanishes at the transition temperature when superconductivity ceases to exist. The BCS theory gives an expression that shows how the gap grows with the strength of the attractive interaction and the (normal phase) single particle density of states at the Fermi level. Furthermore, it describes how the density of states is changed on entering the superconducting state, where there are no electronic states any more at the Fermi level. The energy gap is most directly observed in tunneling experiments and in reflection of microwaves from superconductors.

- BCS theory predicts the dependence of the value of the energy gap Δ at temperature T on the critical temperature $T_c$. The ratio between the value of the energy gap at zero temperature and the value of the superconducting transition temperature (expressed in energy units) takes the universal value

  $$\Delta(T=0) = 1.764 k_B T_c,$$

  independent of material. Near the critical temperature the relation asymptotes to

  $$\Delta(T \to T_c) \approx 3.07 k_B T_c \sqrt{1-(T/T_c)}$$

  which is of the form suggested the previous year by M. J. Buckingham based on the fact that the superconducting phase transition is second order, that the superconducting phase has a mass gap and on Blevins, Gordy and Fairbank's experimental results the previous year on the absorption of millimeter waves by superconducting tin.

- Due to the energy gap, the specific heat of the superconductor is suppressed strongly (exponentially) at low temperatures, there being no thermal excitations left. However, before reaching the transition temperature, the specific heat of the superconductor becomes even higher than that of the normal conductor (measured immediately above the transition) and the ratio of these two values is found to be universally given by 2.5.
- BCS theory correctly predicts the Meissner effect, i.e. the expulsion of a magnetic field from the superconductor and the variation of the penetration depth (the extent of the screening currents flowing below the metal's surface) with temperature. This had been demonstrated experimentally by Walther Meissner and Robert Ochsenfeld in their 1933 article. Ein neuer Effekt bei Eintritt der Supraleitfähigkeit.
- It also describes the variation of the critical magnetic field (above which the superconductor can no longer expel the field but becomes normal conducting) with temperature. BCS theory relates the value of the critical field at zero temperature to the value of the transition temperature and the density of states at the Fermi level.
- In its simplest form, BCS gives the superconducting transition temperature $T_c$ in terms of the electron-phonon coupling potential $V$ and the Debye cutoff energy $E_D$:

$$k_B T_c = 1.13 E_D e^{-1/N(0)V},$$

  where $N(0)$ is the electronic density of states at the Fermi level.
- The BCS theory reproduces the isotope effect, which is the experimental observation that for a given superconducting material, the critical temperature is inversely proportional to the mass of the isotope used in the material. The isotope effect was reported by two groups on 24 March 1950, who discovered it independently working with different mercury isotopes, although a few days before publication they learned of each other's results at the ONR conference in Atlanta. The two groups are Emanuel Maxwell, who published his results in Isotope Effect in the Superconductivity of Mercury and C. A. Reynolds, B. Serin, W. H. Wright, and L. B. Nesbitt who published their results 10 pages later in Superconductivity of Isotopes of Mercury. The choice of isotope ordinarily has little effect on the electrical properties of a material, but does affect the frequency of lattice vibrations. This effect suggests that superconductivity is related to vibrations of the lattice. This is incorporated into BCS theory, where lattice vibrations yield the binding energy of electrons in a Cooper pair.
- Little-Parks experiment - One of the first indications to the importance of the Cooper-pairing principle.

## Wave Function

A wave function in quantum physics is a mathematical description of the quantum state of a system. The wave function is a complex-valued probability amplitude, and the probabilities for the possible results of measurements made on the system can be derived from it. The most common symbols for a wave function are the Greek letters $\psi$ or $\Psi$ (lower-case and capital psi, respectively).

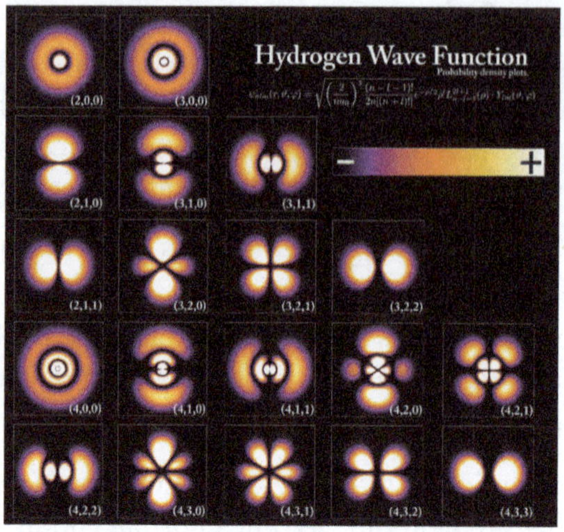


Wavefunctions of the electron of a hydrogen atom at different energies. The brightness at each point represents the probability of finding the electron at that point.

The wave function is a function of the degrees of freedom corresponding to some maximal set of commuting observables. Once such a representation is chosen, the wave function can be derived from the quantum state.

For a given system, the choice of which commuting degrees of freedom to use is not unique, and correspondingly the domain of the wave function is also not unique. For instance it may be taken to be a function of all the position coordinates of the particles over *position space*, or the momenta of all the particles over *momentum space*; the two are related by a Fourier transform. Some particles, like electrons and photons, have nonzero spin, and the wave function for such particles includes spin as an intrinsic, discrete degree of freedom; other discrete variables can also be included, such as isospin. When a system has internal degrees of freedom, the wave function at each point in the continuous degrees of freedom (e.g., a point in space) assigns a complex number for *each* possible value of the discrete degrees of freedom (e.g., z-component of spin) -- these values are often displayed in a column matrix (e.g., a $2 \times 1$ column vector for a non-relativistic electron with spin $\frac{1}{2}$).

According to the superposition principle of quantum mechanics, wave functions can be added together and multiplied by complex numbers to form new wave functions and form a Hilbert space. The inner product between two wave functions is a measure of the overlap between the corresponding physical states, and is used in the foundational probabilistic interpretation of quantum mechanics, the Born rule, relating transition probabilities to inner products. The Schrödinger equation determines how wave functions evolve over time, and a wave function behaves qualitatively like other waves, such as water waves or waves on a string, because the Schrödinger equation is mathematically a type of wave equation. This explains the name "wave function," and gives rise to wave–particle duality. However, the wave function in quantum mechanics describes a kind of physical phenomenon, still open to different interpretations, which fundamentally differs from that of classic mechanical waves.

In Born's statistical interpretation in non-relativistic quantum mechanics, the squared modulus of the wave function, $|\psi|^2$, is a real number interpreted as the probability density of measuring a particle's being detected at a given place - or having a given momentum - at a given time, and possibly having definite values for discrete degrees of freedom. The integral of this quantity, over all the system's degrees of freedom, must be 1 in accordance with the probability interpretation. This general requirement that a wave function must satisfy is called the *normalization condition*. Since the wave function is complex valued, only its relative phase and relative magnitude can be measured -- its value does not, in isolation, tell anything about the magnitudes or directions of measurable observables; one has to apply quantum operators, whose eigenvalues correspond to sets of possible results of measurements, to the wave function $\psi$ and calculate the statistical distributions for measurable quantities.

## Historical Background

In 1905 Einstein postulated the proportionality between the frequency of a photon and its energy, $E = hf$, and in 1916 the corresponding relation between photon momentum and wavelength, $\lambda = h/p$. In 1923, De Broglie was the first to suggest that the relation $\lambda = h/p$, now called the De Broglie relation, holds for *massive* particles, the chief clue being Lorentz invariance, and this can be viewed as the starting point for the modern development of quantum mechanics. The equations represent wave–particle duality for both massless and massive particles.

In the 1920s and 1930s, quantum mechanics was developed using calculus and linear algebra. Those who used the techniques of calculus included Louis de Broglie, Erwin Schrödinger, and others, developing "wave mechanics". Those who applied the methods of linear algebra included Werner Heisenberg, Max Born, and others, developing "matrix mechanics". Schrödinger subsequently showed that the two approaches were equivalent.

In 1926, Schrödinger published the famous wave equation now named after him, indeed the Schrödinger equation, based on classical Conservation of energy using quantum operators and the de Broglie relations such that the solutions of the equation are the wave functions for the quantum system. However, no one was clear on how to *interpret it*. At first, Schrödinger and others thought that wave functions represent particles that are spread out with most of the particle being where the wave function is large. This was shown to be incompatible with the elastic scattering of a wave packet (representing a particle) off a target; it spreads out in all directions. While a scattered particle may scatter in any direction, it does not break up and take off in all directions. In 1926, Born provided the perspective of probability amplitude. This relates calculations of quantum mechanics directly to probabilistic experimental observations. It is accepted as part of the Copenhagen interpretation of quantum mechanics. There are many other interpretations of quantum mechanics. In 1927, Hartree and Fock made the first step in an attempt to solve the $N$-body wave function, and developed the *self-consistency cycle*: an iterative algorithm to approximate the solution. Now it is also known as the Hartree–Fock method. The Slater determinant and permanent (of a matrix) was part of the method, provided by John C. Slater.

Schrödinger did encounter an equation for the wave function that satisfied relativistic energy conservation *before* he published the non-relativistic one, but discarded it as it predicted negative probabilities and negative energies. In 1927, Klein, Gordon and Fock also found it, but incorporated the electromagnetic interaction and proved that it was Lorentz invariant. De Broglie also arrived at the same equation in 1928. This relativistic wave equation is now most commonly known as the Klein–Gordon equation.

In 1927, Pauli phenomenologically found a non-relativistic equation to describe spin-1/2 particles in electromagnetic fields, now called the Pauli equation. Pauli found the wave function was not described by a single complex function of space and time, but needed two complex numbers, which respectively correspond to the spin +1/2 and −1/2 states of the fermion. Soon after in 1928, Dirac found an equation from the first successful unification of special relativity and quantum mechanics applied to the electron, now called the Dirac equation. In this, the wave function is a *spinor* represented by four complex-valued components: two for the electron and two for the electron's antiparticle, the positron. In the non-relativistic limit, the Dirac wave function resembles the Pauli wave function for the electron. Later, other relativistic wave equations were found.

## Wave Functions and Wave Equations in Modern Theories

All these wave equations are of enduring importance. The Schrödinger equation and the Pauli equation are under many circumstances excellent approximations of the relativistic variants. They are considerably easier to solve in practical problems than the relativistic counterparts.

The Klein-Gordon equation and the Dirac equation, while being relativistic, do not represent full reconciliation of quantum mechanics and special relativity. The branch of quantum mechanics where these equations are studied the same way as the Schrödinger equation, often called relativistic quantum mechanics, while very successful, has its limitations and conceptual problems.

Relativity makes it inevitable that the number of particles in a system is not constant. For full reconciliation, quantum field theory is needed. In this theory, the wave equations and the wave functions have their place, but in a somewhat different guise. The main objects of interest are not the wave functions, but rather operators, so called *field operators* (or just fields where "operator" is understood) on the Hilbert space of states. It turns out that the original relativistic wave equations and their solutions are still needed to build the Hilbert space. Moreover, the *free fields operators*, i.e. when interactions are assumed not to exist, turn out to (formally) satisfy the same equation as do the fields (wave functions) in many cases.

Thus the Klein-Gordon equation (spin 0) and the Dirac equation (spin $^1/_2$) in this guise remain in the theory. Higher spin analogues include the Proca equation (spin 1), Rarita–Schwinger equation (spin $^3/_2$), and, more generally, the Bargmann–Wigner equations. For *massless* free fields two examples are the free field Maxwell equation (spin 1) and the free field Einstein equation (spin 2) for the field operators. All of them are essentially a direct consequence of the requirement of Lorentz invariance. Their solutions must transform under Lorentz transformation in a prescribed way, i.e. under a particular representation of the Lorentz group and that together with few other reasonable demands, e.g. the *cluster decomposition principle*, with implications for causality is enough to fix the equations.

It should be emphasized that this applies to free field equations; interactions are not included. If a Lagrangian density (including interactions) is available, then the Lagrangian formalism will yield an equation of motion at the classical level. This equation may be very complex and not amenable to solution. Any solution would refer to a *fixed* number of particles and would not account for the term "interaction" as referred to in these theories, which involves the creation and annihilation of particles and not external potentials as in ordinary "first quantized" quantum theory.

In string theory, the situation remains analogous. For instance, a wave function in momentum space has the role of Fourier expansion coefficient in a general state of a particle (string) with momentum that is not sharply defined.

## Definition (One Spinless Particle in 1d)

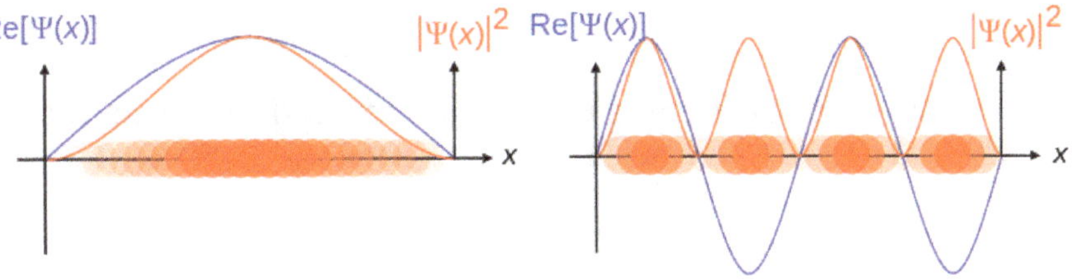


Standing waves for a particle in a box, examples of stationary states.

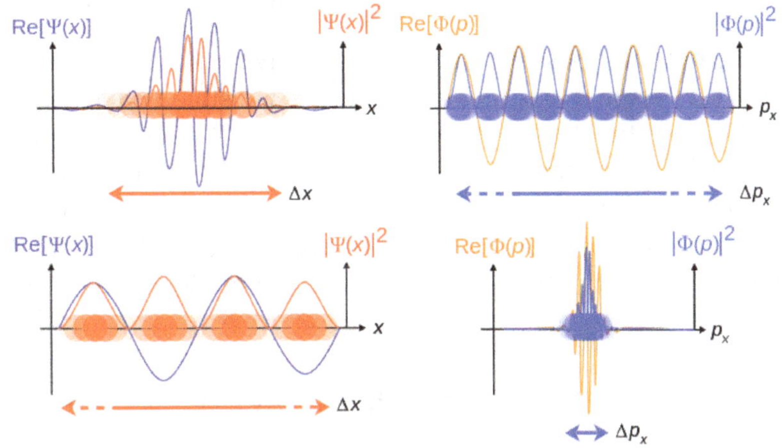


Travelling waves of a free particle. The real parts of position wave function $\Psi(x)$ and momentum wave function $\Phi(p)$, and corresponding probability densities $|\Psi(x)|^2$ and $|\Phi(p)|^2$, for one spin-0 particle in one $x$ or $p$ dimension. The colour opacity of the particles corresponds to the probability density (*not* the wave function) of finding the particle at position $x$ or momentum $p$.

For now, consider the simple case of a non-relativistic single particle, without spin, in one spatial dimension. More general cases are discussed below.

## Position-space Wave Functions

The state of such a particle is completely described by its wave function,

$$\Psi(x,t),$$

where $x$ is position and $t$ is time. This is a complex-valued function of two real variables $x$ and $t$.

For one spinless particle in 1d, if the wave function is interpreted as a probability amplitude, the square modulus of the wave function, the positive real number

$$|\Psi(x,t)|^2 = \Psi(x,t)^* \Psi(x,t) = \rho(x,t),$$

is interpreted as the probability density that the particle is at $x$. The asterisk indicates the complex conjugate. If the particle's position is measured, its location cannot be determined from the wave function, but is described by a probability distribution. The probability that its position $x$ will be in the interval $a \le x \le b$ is the integral of the density over this interval:

$$P_{a\le x\le b}(t) = \int_a^b dx|\Psi(x,t)|^2$$

where t is the time at which the particle was measured. This leads to the normalization condition:

$$\int_{-\infty}^{\infty} \mathrm{dx}|\Psi(\mathrm{x},\mathrm{t})|^2 = 1,$$

because if the particle is measured, there is 100% probability that it will be *somewhere.*

For a given system, the set of all possible normalizable wave functions (at any given time) forms an abstract mathematical vector space, meaning that it is possible to add together different wave functions, and multiply wave functions by complex numbers. Technically, because of the normalization condition, wave functions form a projective space rather than an ordinary vector space. This vector space is infinite-dimensional, because there is no finite set of functions which can be added together in various combinations to create every possible function. Also, it is a Hilbert space, because the inner product of two wave functions $\Psi_1$ and $\Psi_2$ can be defined as the complex number (at time $t$)

$$(\Psi_1, \Psi_2) = \int_{-\infty}^{\infty} \mathrm{dx}\, \Psi_1^*(\mathrm{x},\mathrm{t})\Psi_2(\mathrm{x},\mathrm{t}).$$

More details are given below. Although the inner product of two wave functions is a complex number, the inner product of a wave function $\Psi$ with itself,

$$(\Psi, \Psi) = \|\Psi\|^2,$$

is *always* a positive real number. The number $||\Psi||$ (not $||\Psi||^2$) is called the norm of the wave function $\Psi$, and is not the same as the modulus $|\Psi|$.

If $(\Psi, \Psi) = 1$, then $\Psi$ is normalized. If $\Psi$ is not normalized, then dividing by its norm gives the normalized function $\Psi/||\Psi||$. Two wave functions $\Psi_1$ and $\Psi_2$ are orthogonal if $(\Psi_1, \Psi_2) = 0$. If they are normalized *and* orthogonal, they are orthonormal. Orthogonality (hence also orthonormality) of wave functions is not a necessary condition wave functions must satisfy, but is instructive to consider since this guarantees linear independence of the functions. In a linear combination of orthogonal wave functions $\Psi_n$ we have,

$$\Psi = \sum_{\mathrm{n}} \mathrm{a_n}\Psi_{\mathrm{n}}, \quad \mathrm{a_n} = \frac{(\Psi_{\mathrm{n}}, \Psi)}{(\Psi_{\mathrm{n}}, \Psi_{\mathrm{n}})}$$

If the wave functions $\Psi_n$ were nonorthogonal, the coefficients would be less simple to obtain.

In the Copenhagen interpretation, the modulus squared of the inner product (a complex number) gives a real number

$$|(\Psi_1, \Psi_2)|^2 = \mathrm{P}(\Psi_2 \to \Psi_1),$$

which, assuming both wave functions are normalized, is interpreted as the probability of the wave function $\Psi_2$ "collapsing" to the new wave function $\Psi_1$ upon measurement of an observable, whose eigenvalues are the possible results of the measurement, with $\Psi_1$ being an eigenvector of the resulting eigenvalue. This is the Born rule, and is one of the fundamental postulates of quantum mechanics.

At a particular instant of time, all values of the wave function $\Psi(x, t)$ are components of a vector. There are uncountably infinitely many of them and integration is used in place of summation. In Bra–ket notation, this vector is written

$$|\Psi(t)\rangle = \int dx\Psi(x,t)\,|\,x\rangle$$

and is referred to as a "quantum state vector", or simply "quantum state".There are several advantages to understanding wave functions as representing elements of an abstract vector space:

- All the powerful tools of linear algebra can be used to manipulate and understand wave functions. For example:
  - Linear algebra explains how a vector space can be given a basis, and then any vector in the vector space can be expressed in this basis. This explains the relationship between a wave function in position space and a wave function in momentum space, and suggests that there are other possibilities too.
  - Bra–ket notation can be used to manipulate wave functions.
- The idea that quantum states are vectors in an abstract vector space is completely general in all aspects of quantum mechanics and quantum field theory, whereas the idea that quantum states are complex-valued "wave" functions of space is only true in certain situations.

The time parameter is often suppressed, and will be in the following. The $x$ coordinate is a continuous index. The $|x\rangle$ are the basis vectors, which are orthonormal so their inner product is a delta function;

$$\langle x'\,|\,x\rangle = \delta(x'-x)$$

thus

$$\langle x'\,|\,\Psi\rangle = \int dx\Psi(x)\langle x'\,|\,x\rangle = \Psi(x')$$

and

$$|\Psi\rangle = \int dx\,|\,x\rangle\langle x\,|\,\Psi\rangle = \left(\int dx\,|\,x\rangle\langle x\,|\right)|\,\Psi\rangle$$

which illuminates the identity operator

$$I = \int dx\,|\,x\rangle\langle x\,|.$$

Finding the identity operator in a basis allows the abstract state to be expressed explicitly in a basis, and more (the inner product between two state vectors, and other operators for observables, can be expressed in the basis).

## Momentum-space Wave Functions

The particle also has a wave function in momentum space:

$$\Phi(p,t)$$

where $p$ is the momentum in one dimension, which can be any value from $-\infty$ to $+\infty$, and $t$ is time.

Analogous to the position case, the inner product of two wave functions $\Phi_1(p, t)$ and $\Phi_2(p, t)$ can be defined as:

$$(\Phi_1,\Phi_2)=\int_{-\infty}^{\infty} dp\Phi_1^*(p,t)\Phi_2(p,t).$$

One particular solution to the time-independent Schrödinger equation is

$$\Psi_p(x)=e^{ipx/\hbar},$$

a plane wave, which can be used in the description of a particle with momentum exactly $p$, since it is an eigenfunction of the momentum operator. These functions are not normalizable to unity (they aren't square-integrable), so they are not really elements of physical Hilbert space. The set

$$\{\Psi_p(x,t), -\infty \le p \le \infty\}$$

forms what is called the momentum basis. This "basis" is not a basis in the usual mathematical sense. For one thing, since the functions aren't normalizable, they are instead normalized to a delta function,

$$(\Psi_p,\Psi_{p'})=\delta(p-p').$$

For another thing, though they are linearly independent, there are too many of them (they form an uncountable set) for a basis for physical Hilbert space. They can still be used to express all functions in it using Fourier transforms as described next.

## Relations between Position and Momentum Representations

The $x$ and $p$ representations are

$$|\Psi\rangle = I|\Psi\rangle = \int |x\rangle\langle x|\Psi\rangle dx = \int \Psi(x)|x\rangle dx,$$
$$|\Psi\rangle = I|\Psi\rangle = \int |p\rangle\langle p|\Psi\rangle dp = \int \Phi(p)|p\rangle dp.$$

Now take the projection of the state $\Psi$ onto eigenfunctions of momentum using the last expression in the two equations,

$$\int \Psi(x)\langle p|x\rangle dx = \int \Phi(p')\langle p|p'\rangle dp' = \int \Phi(p')\delta(p-p')dp' = \Phi(p).$$

Then utilizing the known expression for suitably normalized eigenstates of momentum in the position representation solutions of the free Schrödinger equation

$$\langle x | p \rangle = p(x) = \frac{1}{\sqrt{2\pi\hbar}} e^{\frac{i}{\hbar}px} \Rightarrow \langle p | x \rangle = \frac{1}{\sqrt{2\pi\hbar}} e^{-\frac{i}{\hbar}px},$$

one obtains

$$\Phi(p) = \frac{1}{\sqrt{2\pi\hbar}} \int \Psi(x) e^{-\frac{i}{\hbar}px} dx.$$

Likewise, using eigenfunctions of position,

$$\Psi(x) = \frac{1}{\sqrt{2\pi\hbar}} \int \Phi(p) e^{\frac{i}{\hbar}px} dp.$$

The position-space and momentum-space wave functions are thus found to be Fourier transforms of each other. The two wave functions contain the same information, and either one alone is sufficient to calculate any property of the particle. As representatives of elements of abstract physical Hilbert space, whose elements are the possible states of the system under consideration, they represent the same state vector, hence *identical physical states*, but they are not generally equal when viewed as square-integrable functions.

In practice, the position-space wave function is used much more often than the momentum-space wave function. The potential entering the relevant equation (Schrödinger, Dirac, etc.) determines in which basis the description is easiest. For the harmonic oscillator, $x$ and $p$ enter symmetrically, so there it doesn't matter which description one uses. The same equation (modulo constants) results. From this follows, with a little bit of afterthought, a factoid: The solutions to the wave equation of the harmonic oscillator are eigenfunctions of the Fourier transform in $L^2$.

## Definitions (Other Cases)

Following are the general forms of the wave function for systems in higher dimensions and more particles, as well as including other degrees of freedom than position coordinates or momentum components.

## One-particle States in 3d Position Space

The position-space wave function of a single particle without spin in three spatial dimensions is similar to the case of one spatial dimension above:

$$\Psi(\mathbf{r}, t)$$

where r is the position vector in three-dimensional space, and $t$ is time. As always $\Psi(\mathrm{r}, t)$ is a complex-valued function of real variables. As a single vector in Dirac notation

$$|\Psi(t)\rangle = \int d^3\mathbf{r}\, \Psi(\mathbf{r}, t) | \mathbf{r} \rangle$$

All the previous remarks on inner products, momentum space wave functions, Fourier transforms, and so on extend to higher dimensions.

For a particle with spin, ignoring the position degrees of freedom, the wave function is a function of spin only (time is a parameter);

$$\xi(s_z, t)$$

where $s_z$ is the spin projection quantum number along the *z* axis. (The *z* axis is an arbitrary choice; other axes can be used instead if the wave function is transformed appropriately.) The $s_z$ parameter, unlike **r** and *t*, is a *discrete variable*. For example, for a spin-1/2 particle, $s_z$ can only be +1/2 or −1/2, and not any other value. (In general, for spin *s*, $s_z$ can be *s*, *s* − 1, ... , −*s* + 1, −*s*). Inserting each quantum number gives a complex valued function of space and time, there are 2*s* + 1 of them. These can be arranged into a column vector.

$$\xi = \begin{bmatrix} \xi(s,t) \\ \xi(s-1,t) \\ \vdots \\ \xi(-(s-1),t) \\ \xi(-s,t) \end{bmatrix} = \xi(s,t)\begin{bmatrix} 1 \\ 0 \\ \vdots \\ 0 \\ 0 \end{bmatrix} + \xi(s-1,t)\begin{bmatrix} 0 \\ 1 \\ \vdots \\ 0 \\ 0 \end{bmatrix} + \cdots + \xi(-(s-1),t)\begin{bmatrix} 0 \\ 0 \\ \vdots \\ 1 \\ 0 \end{bmatrix} + \xi(-s,t)\begin{bmatrix} 0 \\ 0 \\ \vdots \\ 0 \\ 1 \end{bmatrix}$$

In bra ket notation, these easily arrange into the components of a vector

$$|\xi(t)\rangle = \sum_{s_z=-s}^{s} \xi(s_z, t)\,|\,s_z\rangle$$

The entire vector *ξ* is a solution of the Schrödinger equation (with a suitable Hamiltonian), which unfolds to a coupled system of 2*s* + 1 ordinary differential equations with solutions *ξ*(*s*, *t*), *ξ*(*s* − 1, *t*), ..., *ξ*(−*s*, *t*). The term "spin function" instead of "wave function" is used by some authors. This contrasts the solutions to position space wave functions, the position coordinates being continuous degrees of freedom, because then the Schrödinger equation does take the form of a wave equation.

More generally, for a particle in 3d with any spin, the wave function can be written in "position–spin space" as:

$$\Psi(\mathbf{r}, s_z, t)$$

and these can also be arranged into a column vector

$$\Psi(\mathbf{r}, t) = \begin{bmatrix} \Psi(\mathbf{r}, s, t) \\ \Psi(\mathbf{r}, s-1, t) \\ \vdots \\ \Psi(\mathbf{r}, -(s-1), t) \\ \Psi(\mathbf{r}, -s, t) \end{bmatrix}$$

in which the spin dependence is placed in indexing the entries, and the wave function is a complex vector-valued function of space and time only.

All values of the wave function, not only for discrete but continuous variables also, collect into a single vector

$$|\Psi(t)\rangle = \sum_{s_z} \int d^3\mathbf{r}\Psi(\mathbf{r}, s_z, t)\,|\mathbf{r}, s_z\rangle$$

For a single particle, the tensor product ⊗ of its position state vector $|\psi\rangle$ and spin state vector $|\xi\rangle$ gives the composite position-spin state vector

$$|\psi(t)\rangle \otimes |\xi(t)\rangle = \sum_{s_z} \int d^3\mathbf{r}\psi(\mathbf{r}, t)\xi(s_z, t)\,|\mathbf{r}\rangle \otimes |s_z\rangle$$

with the identifications

$$|\Psi(t)\rangle = |\psi(t)\rangle \otimes |\xi(t)\rangle$$

$$\Psi(\mathbf{r}, s_z, t) = \psi(\mathbf{r}, t)\xi(s_z, t)$$

$$|\mathbf{r}, s_z\rangle = |\mathbf{r}\rangle \otimes |s_z\rangle$$

The tensor product factorization is only possible if the orbital and spin angular momenta of the particle are separable in the Hamiltonian operator underlying the system's dynamics (in other words, the Hamiltonian can be split into the sum of orbital and spin terms). The time dependence can be placed in either factor, and time evolution of each can be studied separately. The factorization is not possible for those interactions where an external field or any space-dependent quantity couples to the spin; examples include a particle in a magnetic field, and spin-orbit coupling.

The preceding discussion is not limited to spin as a discrete variable, the total angular momentum *J* may also be used. Other discrete degrees of freedom, like isospin, can expressed similarly to the case of spin above.

## Many Particle Sates in 3d Position Space

If there are many particles, in general there is only one wave function, not a separate wave function for each particle. The fact that *one* wave function describes *many* particles is what makes quantum entanglement and the EPR paradox possible. The position-space wave function for *N* particles is written:

$$\Psi(\mathbf{r}_1, \mathbf{r}_2 \cdots \mathbf{r}_N, t)$$

where $r_i$ is the position of the *i*th particle in three-dimensional space, and *t* is time. Altogether, this is a complex-valued function of $3N + 1$ real variables.

In quantum mechanics there is a fundamental distinction between *identical particles* and *distinguishable* particles. For example, any two electrons are identical and fundamentally indistinguishable from each other; the laws of physics make it impossible to "stamp an identification number" on a certain electron to keep track of it. This translates to a requirement on the wave function for a system of identical particles:

$$\Psi(\ldots\mathbf{r}_a,\ldots,\mathbf{r}_b,\ldots)=\pm\Psi(\ldots\mathbf{r}_b,\ldots,\mathbf{r}_a,\ldots)$$

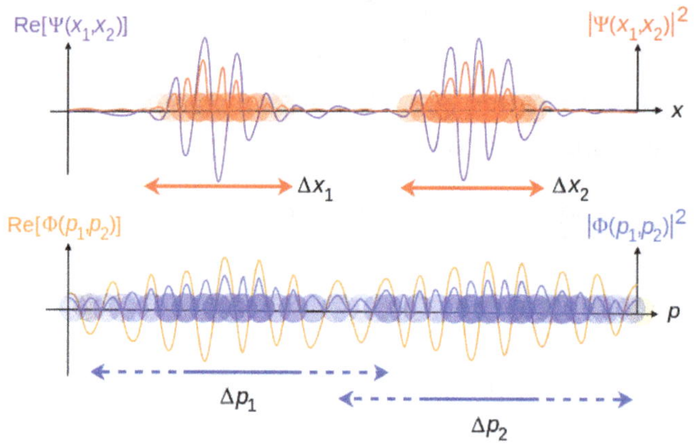


Traveling waves of two free particles, with two of three dimensions suppressed. Top is position space wave function, bottom is momentum space wave function, with corresponding probability densities.

where the + sign occurs if the particles are *all bosons* and – sign if they are *all fermions*. In other words, the wave function is either totally symmetric in the positions of bosons, or totally antisymmetric in the positions of fermions. The physical interchange of particles corresponds to mathematically switching arguments in the wave function. The antisymmetry feature of fermionic wave functions leads to the Pauli principle. Generally, bosonic and fermionic symmetry requirements are the manifestation of particle statistics and are present in other quantum state formalisms.

For *N distinguishable* particles (no two being identical, i.e. no two having the same set of quantum numbers), there is no requirement for the wave function to be either symmetric or antisymmetric.

For a collection of particles, some identical with coordinates $r_1$, $r_2$, ... and others distinguishable $x_1$, $x_2$, ... (not identical with each other, and not identical to the aforementioned identical particles), the wave function is symmetric or antisymmetric in the identical particle coordinates $r_i$ only:

$$\Psi(\ldots\mathbf{r}_a,\ldots,\mathbf{r}_b,\ldots,\mathbf{x}_1,\mathbf{x}_2,\ldots)=\pm\Psi(\ldots\mathbf{r}_b,\ldots,\mathbf{r}_a,\ldots,\mathbf{x}_1,\mathbf{x}_2,\ldots)$$

Again, there is no symmetry requirement for the distinguishable particle coordinates $x_i$.

The wave function for *N* particles each with spin is the complex-valued function

$$\Psi(\mathbf{r}_1,\mathbf{r}_2\cdots\mathbf{r}_N,s_{z1},s_{z2}\cdots s_{zN},t)$$

Accumulating all these components into a single vector,

$$\underbrace{|\Psi\rangle}_{\text{state vector (ket)}} = \underbrace{\overbrace{\sum_{s_{z1},\ldots,s_{zN}}}^{\text{discrete labels}} \overbrace{\int_{R_N} d^3\mathbf{r}_N \cdots \int_{R_1} d^3\mathbf{r}_1}^{\text{continuous labels}}}_{\text{adding up}} \underbrace{\Psi(\mathbf{r}_1,\ldots,\mathbf{r}_N,s_{z1},\ldots,s_{zN})}_{\text{wavefunction (component of state vector along basis state}} \underbrace{|\mathbf{r}_1,\ldots,\mathbf{r}_N,s_{z1},\ldots,s_{zN}\rangle}_{\text{basis state (basis ket)}}.$$

For identical particles, symmetry requirements apply to both position and spin arguments of the wave function so it has the overall correct symmetry.

The formulae for the inner products are integrals over all coordinates or momenta and sums over all spin quantum numbers. For the general case of $N$ particles with spin in 3d,

$$(\Psi_1, \Psi_2) = \sum_{s_{zN}} \cdots \sum_{s_{z2}} \sum_{s_{z1}} \int_{\text{all space}} d^3\mathbf{r}_1 \int_{\text{all space}} d^3\mathbf{r}_2 \cdots \int_{\text{all space}} d^3\mathbf{r}_N \Psi_1^*(\mathbf{r}_1 \cdots \mathbf{r}_N, s_{z1} \cdots s_{zN}, t) \Psi_2(\mathbf{r}_1 \cdots \mathbf{r}_N, s_{z1} \cdots s_{zN}, t)$$

this is altogether $N$ three-dimensional volume integrals and $N$ sums over the spins. The differential volume elements $d^3\mathbf{r}_i$ are also written "$dV_i$" or "$dx_i\, dy_i\, dz_i$".

The multidimensional Fourier transforms of the position or position–spin space wave functions yields momentum or momentum–spin space wave functions.

## Probability Interpretation

For the general case of $N$ particles with spin in 3d, if $\Psi$ is interpreted as a probability amplitude, the probability density is

$$\rho(\mathbf{r}_1 \cdots \mathbf{r}_N, s_{z1} \cdots s_{zN}, t) = \left|\Psi(\mathbf{r}_1 \cdots \mathbf{r}_N, s_{z1} \cdots s_{zN}, t)\right|^2$$

and the probability that particle 1 is in region $R_1$ with spin $s_{z1} = m_1$ *and* particle 2 is in region $R_2$ with spin $s_{z2} = m_2$ etc. at time $t$ is the integral of the probability density over these regions and evaluated at these spin numbers:

$$P_{\mathbf{r}_1 \in R_1, s_{z1}=m_1, \ldots, \mathbf{r}_N \in R_N, s_{zN}=m_N}(t) = \int_{R_1} d^3\mathbf{r}_1 \int_{R_2} d^3\mathbf{r}_2 \cdots \int_{R_N} d^3\mathbf{r}_N \left|\Psi(\mathbf{r}_1 \cdots \mathbf{r}_N, m_1 \cdots m_N, t)\right|^2$$

## Time Dependence

For systems in time-independent potentials, the wave function can always be written as a function of the degrees of freedom multiplied by a time-dependent phase factor, the form of which is given by the Schrödinger equation. For $N$ particles, considering their positions only and suppressing other degrees of freedom,

$$\Psi(\mathbf{r}_1, \mathbf{r}_2, \ldots, \mathbf{r}_N, t) = e^{-iEt/\hbar} \psi(\mathbf{r}_1, \mathbf{r}_2, \ldots, \mathbf{r}_N),$$

where $E$ is the energy eigenvalue of the system corresponding to the eigenstate $\Psi$. Wave functions of this form are called stationary states.

The time dependence of the quantum state and the operators can be placed according to unitary transformations on the operators and states. For any quantum state $|\Psi\rangle$ and operator $O$, in the Schrödinger picture $|\Psi(t)\rangle$ changes with time according to the Schrödinger equation while $O$ is constant. In the Heisenberg picture it is the other way round, $|\Psi\rangle$ is constant while $O(t)$ evolves with time according to the Heisenberg equation of motion. The Dirac (or interaction) picture is intermediate, time dependence is places in both operators and states which evolve according to equations of motion. It is useful primarily in computing S-matrix elements.

## Non-relativistic Examples

The following are solutions to the Schrödinger equation for one nonrelativistic spinless particle.

## Finite Potential Barrier

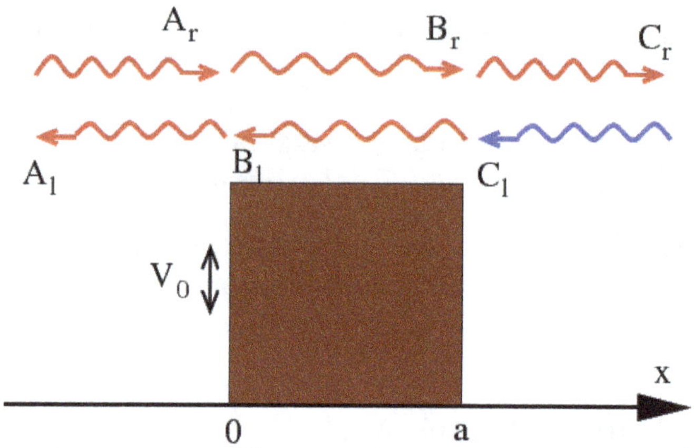


Scattering at a finite potential barrier of height $V_0$. The amplitudes and direction of left and right moving waves are indicated. In red, those waves used for the derivation of the reflection and transmission amplitude. $E > V_0$ for this illustration.

One of most prominent features of the wave mechanics is a possibility for a particle to reach a location with a prohibitive (in classical mechanics) force potential. A common model is the "potential barrier", the one-dimensional case has the potential

$$V(x) = \begin{cases} V_0 & |x| < a \\ 0 & |x| \geq L \end{cases}$$

and the steady-state solutions to the wave equation have the form (for some constants $k$, $\kappa$)

$$\Psi(x) = \begin{cases} A_r e^{ikx} + A_l e^{-ikx} & x < -a, \\ B_r e^{\kappa x} + B_l e^{-\kappa x} & |x| \leq a, \\ C_r e^{ikx} + C_l e^{-ikx} & x > a. \end{cases}$$

Note that these wave functions are not normalized.

The standard interpretation of this is as a stream of particles being fired at the step from the left (the direction of negative $x$): setting $A_r = 1$ corresponds to firing particles singly; the terms containing $A_r$ and $C_r$ signify motion to the right, while $A_l$ and $C_l$ – to the left. Under this beam interpretation, put $C_l = 0$ since no particles are coming from the right. By applying the continuity of wave functions and their derivatives at the boundaries, it is hence possible to determine the constants above.

In a semiconductor crystallite whose radius is smaller than the size of its exciton Bohr radius, the excitons are squeezed, leading to quantum confinement. The energy levels can then be modeled using the particle in a box model in which the energy of different states is dependent on the length of the box.

## Quantum Harmonic Oscillator

The wave functions for the quantum harmonic oscillator can be expressed in terms of Hermite polynomials $H_n$, they are

$$\Psi_n(x) = \sqrt{\frac{1}{2^n n!}} \cdot \left(\frac{m\omega}{\pi\hbar}\right)^{1/4} \cdot e^{-\frac{m\omega x^2}{2\hbar}} \cdot H_n\left(\sqrt{\frac{m\omega}{\hbar}} x\right)$$

where $n = 0,1,2,\ldots.$

## Hydrogen Atom

The wave functions of an electron in a Hydrogen atom are expressed in terms of spherical harmonics and generalized Laguerre polynomials (these are defined differently by different authors.

It is convenient to use spherical coordinates, and the wavefunction can be separated into functions of each coordinate,

$$\Psi(r,\theta,\phi) = R(r) Y_\ell^m(\theta,\phi)$$

where $R$ are radial functions and $Y_\ell^m$ $(\theta, \varphi)$ are spherical harmonics of degree $\ell$ and order $m$. This is the only atom for which the Schrödinger equation has been solved for exactly. Multi-electron atoms require approximative methods. The family of solutions are:

$$\Psi_{n\ell m}(r,\theta,\phi) = \sqrt{\left(\frac{2}{na_0}\right)^3 \frac{(n-\ell-1)!}{2n[(n+\ell)!]}} e^{-r/na_0} \left(\frac{2r}{na_0}\right)^\ell L_{n-\ell-1}^{2\ell+1}\left(\frac{2r}{na_0}\right) \cdot Y_\ell^m(\theta,\phi)$$

where $a_0 = 4\pi\varepsilon_0\hbar^2/m_e e^2$ is the Bohr radius, $L_{n-\ell-1}^{2\ell+1}$ are the generalized Laguerre polynomials of degree $n - \ell - 1$, $n = 1, 2, \ldots$ is the principal quantum number, $\ell = 0, 1, \ldots n - 1$ the azimuthal quantum number, $m = -\ell, -\ell + 1, \ldots, \ell - 1, \ell$ the magnetic quantum number. Hydrogen-like atoms have very similar solutions.

This solution does not take into account the spin of the electron.

In the figure of the hydrogen orbitals, the 19 sub-images are images of wave functions in position space (their norm squared). The wave functions each represent the abstract state characterized by the triple of quantum numbers $(n, l, m)$, in the lower right of each image. These are the principal quantum number, the orbital angular momentum quantum number and the magnetic quantum number. Together with one spin-projection quantum number of the electron, this is a complete set of observables.

The figure can serve to illustrate some further properties of the function spaces of wave functions.

- In this case, the wave functions are square integrable. One can initially take the function space as the space of square integrable functions, usually denoted $L^2$.
- The displayed functions are solutions to the Schrödinger equation. Obviously, not every function in $L^2$ satisfies the Schrödinger equation for the hydrogen atom. The function space is thus a subspace of $L^2$.

- The displayed functions form part of a basis for the function space. To each triple $(n, l, m)$, there corresponds a basis wave function. If spin is taken into account, there are two basis functions for each triple. The function space thus has a countable basis.
- The basis functions are mutually orthonormal.

## Wave Functions and Function Spaces

The concept of function spaces enters naturally in the discussion about wave functions. A function space is a set of functions, usually with some defining requirements on the functions (in the present case that they are square integrable), sometimes with an algebraic structure on the set (in the present case a vector space structure with an inner product), together with a topology on the set. The latter will sparsely be used here, it is only needed to obtain a precise definition of what it means for a subset of a function space to be closed. It will be concluded below that the function space of wave functions is a Hilbert space. This observation is the foundation of the predominant mathematical formulation of quantum mechanics.

## Vector Space Structure

A wave function is an element of a function space partly characterized by the following concrete and abstract descriptions.

- The Schrödinger equation is linear. This means that the solutions to it, wave functions, can be added and multiplied by scalars to form a new solution. The set of solutions to the Schrödinger equation is a vector space.
- The superposition principle of quantum mechanics. If $\Psi$ and $\Phi$ are two states in the abstract space of states of a quantum mechanical system, and $a$ and $b$ are any two complex numbers, then $a\Psi + b\Phi$ is a valid state as well. (Whether the null vector counts as a valid state ("no system present") is a matter of definition. The null vector does *not* at any rate describe the vacuum state in quantum field theory.) The set of allowable states is a vector space.

This similarity is of course not accidental. There are also a distinctions between the spaces to keep in mind.

## Representations

Basic states are characterized by a set of quantum numbers. This is a set of eigenvalues of a maximal set of commuting observables. Physical observables are represented by linear operators, also called observables, on the vectors space. Maximality means that there can be added to the set no further algebraically independent observables that commute with the ones already present. A choice of such a set may be called a choice of representation.

- It is a postulate of quantum mechanics that a physically observable quantity of a system, such as position, momentum, or spin, is represented by a linear Hermitian operator on the state space. The possible outcomes of measurement of the quantity are the eigenvalues of the operator. At a deeper level, most observables, perhaps all, arise as generators of symmetries.

- The physical interpretation is that such a set represents what can – in theory – be simultaneously be measured with arbitrary precision. The Heisenberg uncertainty relation prohibits simultaneous exact measurements of two non-commuting observables.

- The set is non-unique. It may for a one-particle system, for example, be position and spin $z$-projection, $(x, S_z)$, or it may be momentum and spin $y$-projection, $(p, S_y)$. In this case, the operator corresponding to position (a multiplication operator in the position representation) and the operator corresponding to momentum (a differential operator in the position the position representation) do not commute.

- Once a representation is chosen, there is still arbitrariness. It remains to choose a coordinate system. This may, for example, correspond to a choice of $x$, $y$- and $z$-axis, or a choice of curvilinear coordinates as exemplified by the spherical coordinates used for the Hydrogen atomic wave functions. This final choice also fixes a basis in abstract Hilbert space. The basic states are labeled by the quantum numbers corresponding to the maximal set of commuting observables and an appropriate coordinate system.

The abstract states are "abstract" only in that an arbitrary choice necessary for a particular *explicit* description of it is not given. This is the same as saying that no choice of maximal set of commuting observables has been given. This is analogous to a vector space without a specified basis. Wave functions corresponding to a state are accordingly not unique. This non-uniqueness reflects the non-uniqueness in the choice of a maximal set of commuting observables. For one spin particle in one dimension, to a particular state there corresponds two wave functions, $\Psi(x, S_z)$ and $\Psi(p, S_y)$, both describing the *same* state.

- For each choice of maximal commuting sets of observables for the abstract state space, there is a corresponding representation that is associated to a function space of wave functions.

- Between all these different function spaces and the abstract state space, there are one-to-one correspondences (here disregarding normalization and unobservable phase factors), the common denominator here being a particular abstract state. The relationship between the momentum and position space wave functions, for instance, describing the same state is the Fourier transform.

Each choice of representation should be thought of as specifying a unique function space in which wave functions corresponding to that choice of representation lives. This distinction is best kept, even if one could argue that two such function spaces are mathematically equal, e.g. being the set of square integrable functions. One can then think of the function spaces as two distinct copies of that set.

## Inner Product

There is additional algebraic structure on the vector spaces of wave functions and the abstract state space.

- Physically, different wave functions are interpreted to overlap to some degree. A system in a state $\Psi$ that does *not* overlap with a state $\Phi$ cannot be found to be in the state $\Phi$ upon measurement. But if $\Phi_1$, $\Phi_2$, ... overlap $\Psi$ to *some* degree, there is a chance that measurement of a system described by $\Psi$ will be found un states $\Phi_1$, $\Phi_2$, .... Also selection rules are observed apply. These are usually formulated in the preservation of some quantum numbers. This means that certain processes allowable from some perspectives (e.g. energy and momentum conservation) do not occur because the initial and final *total* wave functions don't overlap.

- Mathematically, it turns out that solutions to the Schrödinger equation for particular potentials are orthogonal in some manner, this is usually described by an integral

$$\int \Psi_m^* \Psi_n \mathrm{w} dV = \delta_{nm},$$

where *m*, *n* are (sets of) indices (quantum numbers) labeling different solutions, the strictly positive function *w* is called a weight function, and $\delta_{mn}$ is the Kronecker delta. The integration is taken over all of the relevant space.

This motivates the introduction of an inner product on the vector space of abstract quantum states, compatible with the mathematical observations of above when passing to a representation. It is denoted $(\Psi, \Phi)$, or in the Bra–ket notation $\langle\Psi|\Phi\rangle$. It yields a complex number. With the inner product, the function space is an inner product space. The explicit appearance of the inner product (usually an integral or a sum of integrals) depends on the choice of representation, but the complex number $(\Psi, \Phi)$ does not. Much of the physical interpretation of quantum mechanics stems from the Born rule. It states that the probability *p* of finding upon measurement the state $\Phi$ given the system is in the state $\Psi$ is

$$p = |(\Phi, \Psi)|^2,$$

where $\Phi$ and $\Psi$ are assumed normalized. Consider a scattering experiment. In quantum field theory, if $\Phi_{out}$ describes a state in the "distant future" (an "out state") after interactions between scattering particles have ceased, and $\Psi_{in}$ an "in state" in the "distant past", then the quantities $(\Phi_{out}, \Psi_{in})$, with $\Phi_{out}$ and $\Psi_{in}$ varying over a complete set of in states and out states respectively, is called the S-matrix or scattering matrix. Knowledge of it is, effectively, having *solved* the theory at hand, at least as far as predictions go. Measurable quantities such as decay rates and scattering cross sections are calculable from the S-matrix.

## Hilbert Space

The above observations encapsulate the essence of the function spaces of which wave functions are elements. However the description is not yet complete. There is a further technical requirement on the function space, that of completeness, that allows one to take limits of sequences in the function space, and be ensured that, if the limit exists, it is an element of the function space. A complete inner product space is called a Hilbert space. The property of completeness is crucial in advanced treatments and applications of quantum mechanics. For instance, the existence of projection operators or orthogonal projections relies on the completeness of the space. These projection operators, in turn, are essential for the statement and proof of many useful theorems, e.g. the spectral theorem. It is not very important in introductory quantum mechanics, and technical details and links may be found in footnotes like the one that follows. The space $L^2$ is a Hilbert space, with inner product presented later. The function space of the example of the figure is a subspace of $L^2$. A subspace of a Hilbert space is a Hilbert space if it is closed.

In summary, the set of all possible normalizable wave functions for a system with a particular choice of basis, together with the null vector, constitute a Hilbert space.

Not all functions of interest are elements of some Hilbert space, say $L^2$. The most glaring example is the set of functions $e^{2\pi i p \cdot x/h}$. These are plane wave solutions of the Schrödinger equation for a free particle, but are not normalizable, hence not in $L^2$. But they are nonetheless fundamental for the description. One can, using them, express functions that *are* normalizable using wave packets. They are, in a sense, a basis (but not a Hilbert space basis, nor a Hamel basis) in which wave functions of interest can be expressed. There is also the artifact "normalization to a delta function" that is frequently employed for notational convenience. The delta functions themselves aren't square integrable either.

The above description of the function space containing the wave functions is mostly mathematically motivated. The function spaces are, due to completeness, very *large* in a certain sense. Not all functions are realistic descriptions of any physical system. For instance, in the function space $L^2$ one can find the function that takes on the value 0 for all rational numbers and $-i$ for the irrationals in the interval [0, 1]. This *is* square integrable, but can hardly represent a physical state.

## Common Hilbert Spaces

While the space of solutions as a whole is a Hilbert space there are many other Hilbert spaces that commonly occur as ingredients.

- Square integrable complex valued functions on the interval $[0, 2\pi]$. The set $\{e^{int}/2\pi, n \in \mathbb{Z}\}$ is a Hilbert space basis, i.e. a maximal orthonormal set.
- The Fourier transform takes functions in the above space to elements of $l^2(\mathbb{Z})$, the space of *square summable* functions $\mathbb{Z} \to \mathbb{C}$. The latter space is a Hilbert space and the Fourier transform is an isomorphism of Hilbert spaces. Its basis is $\{e_i, i \in \mathbb{Z}\}$ with $e_i(j) = \delta_{ij}, i, j \in \mathbb{Z}$.
- The most basic example of spanning polynomials is in the space of square integrable functions on the interval $[-1, 1]$ for which the Legendre polynomials is a Hilbert space basis (complete orthonormal set).
- The square integrable functions on the unit sphere $S^2$ is a Hilbert space. The basis functions in this case are the spherical harmonics. The Legendre polynomials are ingredients in the spherical harmonics. Most problems with rotational symmetry will have "the same" (known) solution with respect to that symmetry, so the original problem is reduced to a problem of lower dimensionality.
- The associated Laguerre polynomials appear in the hydrogenic wave function problem after factoring out the spherical harmonics. These span the Hilbert space of square integrable functions on the semi-infinite interval $[0, \infty)$.

More generally, one may consider a unified treatment of all second order polynomial solutions to the Sturm–Liouville equations in the setting of Hilbert space. These include the Legendre and Laguerre polynomials as well as Chebyshev polynomials, Jacobi polynomials and Hermite polynomials. All of these actually appear in physical problems, the latter ones in the harmonic oscillator, and what is otherwise a bewildering maze of properties of special functions becomes an organized body of facts.

There occurs also finite-dimensional Hilbert spaces. The space $\mathbb{C}^n$ is a Hilbert space of dimension $n$. The inner product is the standard inner product on these spaces. In it, the "spin part" of a single particle wave function resides.

- In the non-relativistic description of an electron one has $n = 2$ and the total wave function is a solution of the Pauli equation.
- In the corresponding relativistic treatment, $n = 4$ and the wave function solves the Dirac equation.

With more particles, the situations is more complicated. One has to employ tensor products and use representation theory of the symmetry groups involved (the rotation group and the Lorentz group respectively) to extract from the tensor product the spaces in which the (total) spin wave functions reside. (Further problems arise in the relativistic case unless the particles are free.) Corresponding remarks apply to the concept of isospin, for which the symmetry group is SU(2). The models of the nuclear forces of the sixties used the symmetry group SU(3). In this case as well, the part of the wave functions corresponding to the inner symmetries reside in some $\mathbb{C}^n$ or subspaces of tensor products of such spaces.

- In quantum field theory the underlying Hilbert space is Fock space. It is built from free single-particle states, i.e. wave functions when a representation is chosen, and can accommodate any finite, not necessarily constant in time, number of particles. The interesting (or rather the *tractable*) dynamics lies not in the wave functions but in the field operators that are operators acting on Fock space. Thus the Heisenberg picture is the most common choice (constant states, time varying operators).

Due to the infinite-dimensional nature of the system, the appropriate mathematical tools are objects of study in functional analysis.

## Simplified Description

Not all introductory textbooks take the long route and introduce the full Hilbert space machinery, but the focus is on the non-relativistic Schrödinger equation in position representation for certain standard potentials. The following constraints on the wave function are sometimes explicitly formulated for the calculations and physical interpretation to make sense:

- The wave function must be square integrable. This is motivated by the Copenhagen interpretation of the wave function as a probability amplitude.
- It must be everywhere continuous and everywhere continuously differentiable. This is motivated by the appearance of the Schrödinger equation for most physically reasonable potentials.

It is possible to relax these conditions somewhat for special purposes. If these requirements are not met, it is not possible to interpret the wave function as a probability amplitude.

This does not alter the structure of the Hilbert space that these particular wave functions inhabit, but it should be pointed out that the subspace of the square-integrable functions $L^2$, which is a Hilbert space, satisfying the second requirement *is not closed* in $L^2$, hence not a Hilbert space in itself. The functions that does not meet the requirements are still needed for both technical and practical reasons.

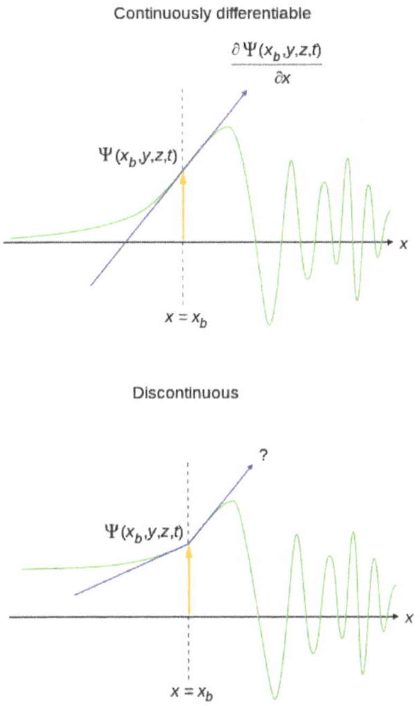


Continuity of the wave function and its first spatial derivative (in the $x$ direction, $y$ and $z$ coordinates not shown), at some time $t$.

## More on Wave Functions and Abstract State Space

As has been demonstrated, the set of all possible wave functions in some representation for a system constitute an in general infinite-dimensional Hilbert space. Due to the multiple possible choices of representation basis, these Hilbert spaces are not unique. One therefore talks about an abstract Hilbert space, state space, where the choice of representation and basis is left undetermined. Specifically, each state is represented as an abstract vector in state space. A quantum state $|\Psi\rangle$ in any representation is generally expressed as a vector

$$|\Psi\rangle = \sum_{\alpha} \int d^{m}\omega \Psi(\alpha, \omega, t) |\alpha, \omega\rangle$$

where $\alpha = (\alpha_1, \alpha_2, ..., \alpha_n)$ are (dimensionless) discrete quantum numbers, and $\omega = (\omega_1, \omega_2, ..., \omega_m)$ are continuous variables (not necessarily dimensionless). All of them index the components of the vector, and $|\alpha, \omega\rangle$ are the basis vectors in this representation. All $\alpha$ are in an $n$-dimensional set $A = A_1 \times A_2 \times ... A_n$ where each $A_i$ is the set of allowed values for $\alpha_i$, likewise all $\omega$ are in an $m$-dimensional "volume" $\Omega \subseteq \mathbb{R}^m$ where $\Omega = \Omega_1 \times \Omega_2 \times ... \Omega_m$ and each $\Omega_i \subseteq \mathbb{R}$ is the set of allowed values for $\omega_i$, a subset of the real numbers $\mathbb{R}$. For generality $n$ and $m$ are not necessarily equal.

For example, for a single particle in 3d with spin $s$, neglecting other degrees of freedom, using Cartesian coordinates, we could take $\alpha = (s_z)$ for the spin quantum number of the particle along the z direction, and $\omega = (x, y, z)$ for the particle's position coordinates. Here $A = \{-s, -s + 1, ..., s - 1, s\}$ is the set of allowed spin quantum numbers and $\Omega = \mathbb{R}^3$ is the set of all possible particle positions throughout 3d position space. An alternative choice is $\alpha = (s_y)$ for the spin quantum number along the y direction and $\omega = (p_x, p_y, p_z)$ for the particle's momentum components. In this case $A$ and $\Omega$ are the same.

Then, a component $\Psi(\alpha, \omega, t)$ of the vector $|\Psi\rangle$ is referred to as the "wave function" of the system.

When interpreted as a probability amplitude (non-relativistic systems with constant number of particles), the probability density of finding the system at $\alpha$, $\omega$ is

$$\rho = |\Psi(\alpha, \omega, t)|^2$$

The probability of finding system with $\alpha$ in some or all possible discrete-variable configurations, $D \subseteq A$, and $\omega$ in some or all possible continuous-variable configurations, $C \subseteq \Omega$, is the sum and integral over the density,

$$P = \sum_{\alpha \in D} \int_C \rho d^m \omega$$

where $d^m\omega = d\omega_1 d\omega_2 ... d\omega_m$ is a "differential volume element" in the continuous degrees of freedom. Since the sum of all probabilities must be 1, the normalization condition

$$1 = \sum_{\alpha \in A} \int_\Omega \rho d^m \omega$$

must hold at all times during the evolution of the system.

The normalization condition requires $\rho\, d^m\omega$ to be dimensionless, by dimensional analysis $\Psi$ must have the same units as $(\omega_1 \omega_2 ... \omega_m)^{-1/2}$.

## Ontology

Whether the wave function really exists, and what it represents, are major questions in the interpretation of quantum mechanics. Many famous physicists of a previous generation puzzled over this problem, such as Schrödinger, Einstein and Bohr. Some advocate formulations or variants of the Copenhagen interpretation (e.g. Bohr, Wigner and von Neumann) while others, such as Wheeler or Jaynes, take the more classical approach and regard the wave function as representing information in the mind of the observer, i.e. a measure of our knowledge of reality. Some, including Schrödinger, Bohm and Everett and others, argued that the wave function must have an objective, physical existence. Einstein thought that a complete description of physical reality should refer directly to physical space and time, as distinct from the wave function, which refers to an abstract mathematical space.

## BCS Wave Function

In Bardeen-Cooper-Schrieffer (BCS) theory of superconductivity the ground state wave function is postulated to be

$$|\psi_{BCS}> = \prod_{\vec{k}} \left( u_{\vec{k}} + v_{\vec{k}} C^*_{\vec{k}\uparrow} C^+_{-\vec{k}\downarrow} \right) |0>$$

where the coefficients $u_{\vec{k}}$ and $v_{\vec{k}}$ are to be determined variationally by minimizing the total energy. By requiring that the BCS ground state be normalized,

$$<\psi_{BCS}|\psi_{BCS}>=1$$

we find that

$$|u_{\vec{k}}|^2+|v_{\vec{k}}|^2=1$$

as shown in the following. Writing for $|\psi_{BCS}>$, we have

$$<\psi_{BCS}|\psi_{BCS}>=<0\left|\prod_{\vec{k}}\left(u_{\vec{k}}^*+v_{\vec{k}}^*C_{-\vec{k}\downarrow}C_{\vec{k}\uparrow}\right)\prod_{\vec{\ell}}\left(u_{\vec{\ell}}+v_{\vec{\ell}}C_{\vec{k}\uparrow}^+C_{-\vec{k}\downarrow}^+\right)\right|0>$$

$\vec{k}\neq\vec{\ell}$ terms will not contribute. Writing out the remaining terms

$$=<0\left|\prod_{\vec{k}}\left(u_{\vec{k}}^*+v_{\vec{k}}^*C_{-\vec{k}\downarrow}C_{\vec{k}\uparrow}\right)\left(u_{\vec{k}}+v_{\vec{k}}C_{\vec{k}\uparrow}C_{-\vec{k}\downarrow}\right)\right|0>$$

$$=\langle 0\left|\prod_{\vec{k}}(|u_{\vec{k}}|^2+v_{\vec{k}}^*u_{\vec{k}}C_{-\vec{k}\downarrow}C_{\vec{k}\uparrow}+u_{\vec{k}}^*v_{\vec{k}}C_{\vec{k}\uparrow}^+C_{-\vec{k}\downarrow}^++|v_{\vec{k}}|^2C_{-\vec{k}\downarrow}C_{\vec{k}\uparrow}C_{\vec{k}\uparrow}^+C_{-\vec{k}\downarrow}^+)\right|0\rangle$$

Note that the second and the third terms in the last equation vanish because they either create or destroy a pair. Now we have

$$<\psi_{BCS}|\psi_{BCS}>=<0\left|\prod_{\vec{k}}\left(|u_{\vec{k}}|^2+|v_{\vec{k}}|^2C_{-\vec{k}|}C_{\vec{k}|}C_{\vec{k}|}^+C_{-\vec{k}|}^+\right)\right|0>$$

$$=\prod_{k}\left(|u_{\vec{k}}|^2+|v_{\vec{k}}|^2\right)$$

So that

$$\langle\psi_{BCS}|\psi_{BCS}\rangle=\prod_{\vec{k}}|u_{\vec{k}}|^2+|v_{\vec{k}}|^2$$

Or,

$$<\psi_{BCS}|\psi_{BCS}>=1$$

That is, the BCS wave function is normalized to one if we require that

$$|u_{\vec{k}}|^2+|v_{\vec{k}}|^2=1$$

for each $\vec{k}$.

## Number of Particles in the BCS Wave Function

To get an idea of how many electrons can be associated with the BCS wave function, we consider the total electron number operator $\hat{N}$,

$$\hat{N} = \sum_{\vec{k}\sigma} n_{\vec{k}\uparrow} + n_{\vec{k}\downarrow} = 2n_{\vec{k}\uparrow}$$

Where $n_{\vec{k}\uparrow} = C^{+}_{\vec{k}\uparrow} C_{\vec{k}\uparrow}$ and $n_{\vec{k}\downarrow} = C^{+}_{\vec{k}\downarrow} C_{\vec{k}\downarrow}$, and in the BCS wave function the number of electrons with up spin is equal to the number of electrons with down spin. The mean number of electrons can be written as

$$\bar{N} = 2\sum_{\vec{k}} < \psi \left| n_{\vec{k}\uparrow} \right| \psi_{BCS} >$$

$$2\sum_{\vec{k}} < \psi \left| C^{+}_{\vec{k}\downarrow} C_{\vec{k}\downarrow} \right| \psi_{BCS} >$$

$$= 2\sum_{k} < 0 \left| \prod_{\vec{\ell}} \left( u^{*}_{\vec{\ell}} + v^{*}_{\vec{\ell}} C_{-\vec{\ell}\downarrow} C_{\vec{\ell}\uparrow} \right) C^{+}_{\vec{k}\uparrow} C_{\vec{k}\uparrow} \left( \prod_{\vec{\ell}'} (u_{\vec{\ell}'} + v_{\vec{\ell}'} C^{+}_{\vec{\ell}'\uparrow} C^{+}_{-\vec{\ell}'\downarrow}) \right) \right| 0 >$$

Only $\vec{\ell} = \vec{\ell}'$ will have any contribution. Writing out the $\vec{\ell} = \vec{k}$ term separately, we have

$$= 2\sum_{\vec{k}} < 0 \left| \sum_{\vec{\ell}} \left( u^{*}_{\vec{k}} + v^{*}_{\vec{k}} C_{-\vec{k}\downarrow} C_{\vec{k}\uparrow} \right) C^{+}_{\vec{k}\uparrow} C_{\vec{k}\uparrow} \left( u_{\vec{k}} + v_{\vec{k}} C^{+}_{\vec{k}\uparrow} C^{+}_{-\vec{k}\uparrow} \right) \right.$$

$$\prod_{\vec{\ell} \neq \vec{k}} \left( u^{*}_{\vec{\ell}} + v^{*}_{\vec{\ell}} C_{-\vec{\ell}\downarrow} C_{\vec{\ell}\uparrow} \right) \left( u_{\vec{\ell}} + v_{\vec{\ell}} C^{+}_{\vec{\ell}\uparrow} C^{+}_{-\vec{\ell}\uparrow} \right)$$

We can evaluate the term involving each pair separately. Consider $\ell \neq k$ term after multiplication

$$\left( u^{*}_{\vec{\ell}} u_{\vec{\ell}} + u^{*}_{\vec{\ell}} v_{\vec{\ell}} C^{+}_{\vec{\ell}\uparrow} C^{+}_{-\vec{\ell}\downarrow} + u_{\vec{\ell}} v^{*}_{\vec{\ell}} C_{-\vec{\ell}\downarrow} C_{\vec{\ell}\uparrow} + v^{*}_{\vec{\ell}} v_{\vec{\ell}} C_{-\vec{\ell}\downarrow} C_{\vec{\ell}\downarrow} C^{+}_{\vec{\ell}\uparrow} C^{+}_{-\vec{\ell}\downarrow} \right)$$

A simple inspection shows that the second and the third terms vanish while the last term yields $\left| v_{\vec{\ell}} \right|^2$ so that the $\vec{\ell} \neq \vec{k}$ term when evaluated between the vacuum state yields

$$\left| u_{\vec{\ell}} \right|^2 + \left| v_{\vec{\ell}} \right|^2 = 1$$

Next we consider $\ell = k$ term,

$$= u_{\vec{k}} v^{*}_{\vec{k}} C^{+}_{\vec{k}\uparrow} C_{\vec{k}\uparrow} + v_{\vec{k}} u^{*}_{\vec{k}} C^{+}_{\vec{k}\uparrow} C_{\vec{k}\uparrow} C^{+}_{\vec{k}\uparrow} C^{+}_{-\vec{k}\downarrow} + u_{\vec{k}} v^{*}_{\vec{k}} C_{-\vec{k}\downarrow} C_{\vec{k}\downarrow} C^{+}_{\vec{k}\downarrow} C_{\vec{k}\uparrow} + v^{*}_{\vec{k}} v_{\vec{k}} C_{-\vec{k}\downarrow} C_{\vec{k}\uparrow} C^{+}_{\vec{k}\uparrow} C_{\vec{k}\uparrow} C^{+}_{\vec{k}\uparrow} C^{+}_{-\vec{k}\downarrow}$$

$$= \left| v_{\vec{k}} \right|^2$$

because (i) the first term vanishes since $C_{\vec{k}\uparrow} \left| 0 \right> = 0$, (ii) the second and the third terms involving $C^{+}_{\vec{k}\uparrow} C_{\vec{k}\uparrow}$ do not affect the particle number, which would be changed by the term on the right, (iii) in the fourth term a pair is created on both sides. Hence, the average number of particles contained in the BCS wave function is equal to

$$\bar{N} = 2\sum_{\vec{k}} |v_{\vec{k}}|^2$$

Similarly, we can evaluate $<\psi_{BCS}|\hat{N}^2|\psi_{BCS}>$ for finding out the root mean square (rms) fluctuation,

$$<\psi_{BCS}|\hat{N}^2|\psi_{BCS}> = 4<\psi_{BCS}|\sum_{\vec{k}}\left(C^{+}_{\vec{k}|}C_{\vec{k}|}\right)\sum_{\vec{k}}\left(C^{+}_{\vec{k}|}C_{\vec{k}|}\right)|\psi_{BCS}>$$

which is found to be

$$<\psi_{BCS}|\hat{N}^2|\psi_{BCS}> = 4\sum_{k}|u_k|^2|v_k|^2$$

Therefore,

$$\begin{aligned}<\delta N^2> &= <N^2> - <N>^2 \\ &= 4\sum_{k}|v_k|^2\left(|u_k|^2 - |v_k|^2\right)\end{aligned}$$

## BCS Wave Function in Terms of 2m-particle States

The fact that the BCS wave function is normalized to one does not imply that it contains one particle. In fact, it turns out that the number of particles is not fixed by normalizing the BCS wave function.

To understand further the nature of the BCS wave function

$$|\psi_{BCS}> = \prod_{\vec{k}}\left(u_{\vec{k}} + v_{\vec{k}}C^{+}_{\vec{k}\uparrow}C^{+}_{-\vec{k}\downarrow}\right)|0>$$

let us expand it assuming only two distinct $\vec{k}$ values, namely $\vec{k_1}$ and $\vec{k_2}$, and their negative counterparts,

$$\begin{aligned}&= \left(u_{\vec{k}_1} + v_{\vec{k}_1}C^{+}_{\vec{k}_1\uparrow}C^{+}_{-k_1\downarrow}\right)\left(u_{\vec{k}_2} + v_{\vec{k}_2}C^{+}_{\vec{k}_2\uparrow}C^{+}_{-\vec{k}_2\downarrow}\right)\ldots|0> \\ &= (u_{\vec{k}_1}u_{\vec{k}_2}\ldots + u_{\vec{k}_1}v_{\vec{k}_2}C^{+}_{\vec{k}_1\uparrow}C^{+}_{-k_1\downarrow} + v_{\vec{k}_1}v_{\vec{k}_2}C^{+}_{\vec{k}_2\uparrow}C^{+}_{-k_2\downarrow} + v_{\vec{k}_1}v_{\vec{k}_2}C^{+}_{\vec{k}_1\uparrow}C^{+}_{-k_1\downarrow}C^{+}_{\vec{k}_2\uparrow}C^{+}_{-k_2\downarrow})\ldots|0>\end{aligned}$$

From the expansion of the BCS wave function for two distinct $\vec{k}$ values, we see that the $|\psi_{BCS}>$ is nothing but the sum of all $2m$ particle states where $m$ is an integer. Thus, we can rewrite the BCS ground state as

$$|\psi_{BCS}> = \sum_{m} A_{2m}|\psi_{2m}>$$

where $A_{2m}$ is an expansion coefficient and $|\psi_{2m}>$ is the ground state wave function for the $2m$ particles. We can get an expression for $|\psi_{2m}>$ in terms of $|\psi_{BCS}>$ by writing the expansion coefficients as

$$A_{2m} = |A_{2m}|e^{i2m\phi}$$

so that

$$|\psi^{\phi}_{BCS}> = \sum_m |A_{2m}|\, e^{i2m\phi} |\psi_{2m}>$$

In the above, each $2m$ state has been multiplied by a phase factor $e^{i2m\phi}$, which leads to

$$|\psi^{\phi}_{BCS}> = \prod_{\vec{k}} \left(u_{\vec{k}} + v_{\vec{k}} e^{2i\phi} C^{+}_{\vec{k}\uparrow} C^{+}_{-\vec{k}\downarrow}\right)|0>$$

Multiplying Eq. (eq:bcs-wf-phase) by $e^{-i2m'\phi}$ and integrating over $\phi$, we get

$$\int_0^{2\pi} e^{-2im'\phi} |\psi^{\phi}_{BCS}> d\phi = \sum_m |A_{2m}| \int_0^{2\pi} e^{-2im'\phi}\, e^{2im\phi} |\psi_{2m}> d\phi$$

Using

$$\frac{1}{2\pi}\int_0^{2\pi} e^{i\phi(m-m')} d\phi = \delta_{mm'}$$

we get

$$\int_0^{2\pi} e^{-2im'\phi} |\psi^{\phi}_{BCS}> d\phi' = \sum_m |A_{2m}|\, 2\pi\delta_{mm'} |\psi_{2m}>$$
$$= 2\pi A_{2m'} |\psi_{2m'}>$$

Changing $m'$ $to\, m$ on both sides of the equation, we have

$$|\psi_{2m}> = \frac{1}{2\pi |A_{2m}|} \int_0^{2\pi} e^{-2im\phi}\; |\psi^{\phi}_{BCS}> d\,\phi$$

Now the probability amplitude of $2m$ particle states can be obtained after noting that

$$<\psi_{2m}|\psi_{2m}> = 1$$

Hence,

$$1 = \frac{1}{4\pi |A_{2m}|^2} \int_0^{2\pi} d\phi' \int_0^{2\pi} d\phi e^{-2im\phi}\, e^{2im\phi}\; <\psi^{\phi}_{BCS}|\psi^{\phi}_{BCS}>$$

Next we simplify $<\psi^{\phi'}_{BCS}|\psi^{\phi}_{BCS}>$ by substituting the expression given in Eq. (eq:bcs-wf-phi),

$$<\psi^{\phi'}_{BCS}|\psi^{\phi}_{BCS}> = <0|\prod_{\vec{k}\vec{k}'} (u^{*}_{\vec{k}'} + v_{k'} e^{-2i\phi'} C_{-\vec{k}'\downarrow} C_{\vec{k}'\uparrow})(u_k + v_k e^{2i\phi} C^{+}_{\vec{k}\uparrow} C^{+}_{-\vec{k}\downarrow})|0>$$

Only $\vec{k} = \vec{k}'$ terms survive,

$$< \psi_{BCS}^{\phi'} \left| \psi_{BCS}^{\phi} \right> = < 0 \left| \prod_{\vec{k}} (\left| u_{\vec{k}} \right|^2 + \left| v_{\vec{k}} \right|^2 e^{2i(\phi-\phi')} C_{-\vec{k}|} C_{\vec{k}|} C_{\vec{k}|}^{+} C_{-k|}^{+}) \right| 0 >$$

Using the anti-commutation rules $[C_{k\sigma,} C_{k'\sigma'}^{+}]_{+} = \delta_{kk'} \delta_{\sigma\sigma'}$

$$[C_{k\sigma,} C_{k'\sigma'}]_{+} = [C_{k\sigma,}^{+} C_{k'\sigma'}^{+}] = 0$$

we notice that $C_{-\vec{k}\downarrow} C_{\vec{k}\uparrow} C_{\vec{k}\uparrow}^{+} C_{-\vec{k}\downarrow}^{+}$ can be simplified into

$$\left(1 - n_{-\vec{k}\downarrow} - n_{\vec{k}\uparrow} + n_{-\vec{k}\downarrow} n_{\vec{k}\uparrow}\right)$$

As the number operators operate over the vacuum state, $n_{k|}$ and $n_{k|}$ have zero eigenvalues. Thus

$$< \psi_{BCS}^{\phi'} \left| \psi_{BCS}^{\phi} \right> = \left\langle 0 \left| \prod_{\vec{k}} \left| u_k \right|^2 + \left| u_k \right|^2 e^{2i(\phi-\phi')} \right| 0 \right>$$

Collecting all the terms, we get for the probability amplitude of the $2m$ -particle state and integrating over $\phi'$ , we get

$$\left| A_{2m} \right|^2 = \frac{1}{2\pi} \int_0^{2\pi} d\phi e^{i2m\phi} \prod_{\vec{k}} (\left| u_k \right|^2 + \left| v_k \right|^2 e^{2i\phi})$$

The pair amplitude is peaked at the average number of particles and its width is proportional to the square root of the average number of particles.

## References


- Bohr, N. (1985). J. Kalckar, ed. Niels Bohr - Collected Works: Foundations of Quantum Physics I (1926 - 1932). 6. Amsterdam: North Holland. ISBN 9780444532893
- Nave, Carl R. (2006). "Cooper Pairs". Hyperphysics. Dept. of Physics and Astronomy, Georgia State Univ. Retrieved 2008-07-24
- Cooper, Leon N. (1956). "Bound electron pairs in a degenerate Fermi gas". Physical Review. 104 (4): 1189–1190. Bibcode:1956PhRv..104.1189C. doi:10.1103/PhysRev.104.1189
- Young, H. D.; Freedman, R. A. (2008). Pearson, ed. Sears' and Zemansky's University Physics (12th ed.). Addison-Wesley. ISBN 978-0-321-50130-1
- Kadin, Alan M. (2005). "Spatial Structure of the Cooper Pair". Journal of Superconductivity and Novel Magnetism. 20 (4): 285. arXiv:cond-mat/0510279. doi:10.1007/s10948-006-0198-z
- Nave, Carl R. (2006). "The BCS Theory of Superconductivity". Hyperphysics. Dept. of Physics and Astronomy, Georgia State Univ. Retrieved 2008-07-24
- Born, M. (1926a). "Zur Quantenmechanik der Stoßvorgange". Z. Phys. 37: 863–867. Bibcode:1926ZPhy...37..863B. doi:10.1007/bf01397477
- Jaynes, E. T. (2003). G. Larry Bretthorst, ed. Probability Theory: The Logic of Science. Cambridge University Press. ISBN 978-0-521 59271-0
- Bardeen, J.; Cooper, L. N.; Schrieffer, J. R. (April 1957). "Microscopic Theory of Superconductivity". Physical Review. 106 (1): 162–164. Bibcode:1957PhRv..106..162B. doi:10.1103/PhysRev.106.162. Retrieved May 3, 2012

- Dirac, P. A. M. (1939). "A new notation for quantum mechanics". Mathematical Proceedings of the Cambridge Philosophical Society. 35 (3): 416–418. Bibcode:1939PCPS...35..416D. doi:10.1017/S0305004100021162
- Dirac, P. A. M. (1982). The principles of quantum mechanics. The international series on monographs on physics (4th ed.). Oxford University Press. ISBN 0 19 852011 5
- Arons, A. B.; Peppard, M. B. (1965). "Einstein's proposal of the photon concept: A translation of the Annalen der Physik paper of 1905" (PDF). American Journal of Physics. 33 (5): 367. Bibcode:1965AmJPh..33..367A. doi:10.1119/1.1971542
- London, F. (September 1948). "On the Problem of the Molecular Theory of Superconductivity". Physical Review. 74 (5): 562–573. Bibcode:1948PhRv...74..562L. doi:10.1103/PhysRev.74.562. Retrieved March 3, 2012
- Pauli, Wolfgang (1927). "Zur Quantenmechanik des magnetischen Elektrons". Zeitschrift für Physik (in German). 43. Bibcode:1927ZPhy...43..601P. doi:10.1007/bf01397326

# 5

# Superconductivity: Tunneling and Energy Gap

Tunnel junctions are barriers between two electrically conducting materials. Time-dependent perturbation theory can accurately calculate between different tunneling junctions. The aspects elucidated in this chapter are of vital importance, and provide a better understanding of superconductivity.

## Tunnel Junction

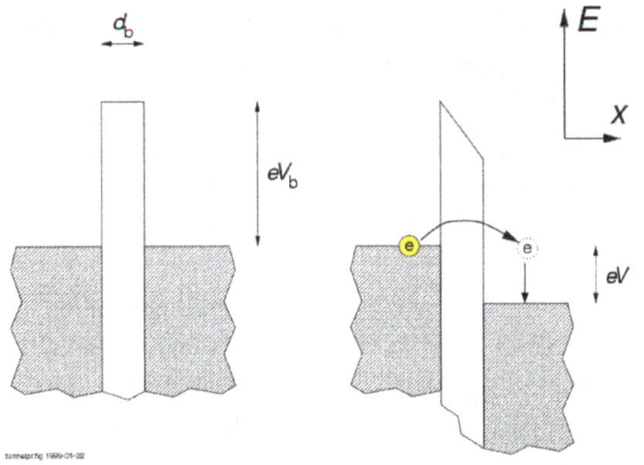


Schematic representation of an electron tunneling through a barrier.

In electronics, a tunnel junction is a barrier, such as a thin insulating layer or electric potential, between two electrically conducting materials. Electrons (or quasiparticles) pass through the barrier by the process of quantum tunneling. Classically, the electron has zero probability of passing through the barrier. However, according to quantum mechanics, the electron has a non-zero wave amplitude in the barrier, and hence it has some probability of passing through the barrier. Tunnel junctions serve a variety of different purposes.

- In multijunction photovoltaic cells, tunnel junctions form the connections between consecutive p-n junctions. They function as an ohmic electrical contact in the middle of a semiconductor device.
- In magnetic tunnel junctions, electrons tunnel through a thin insulating barrier from one magnetic material to another. This can serve as a basis for a magnetic detector.
- In superconducting tunnel junctions, two superconducting electrodes are separated by a non-superconducting barrier. Cooper pairs carry the supercurrent through the barrier by quantum tunneling, a phenomenon known as the Josephson effect. This setup can form the basis for extremely sensitive magnetometers, known as SQUIDs, as well as many other devices.

- In tunnel diodes, a diode allows the tunneling of electrons for certain voltages. This allows them to be used for generating high-frequency signals.

## Superconducting Tunnel Junction

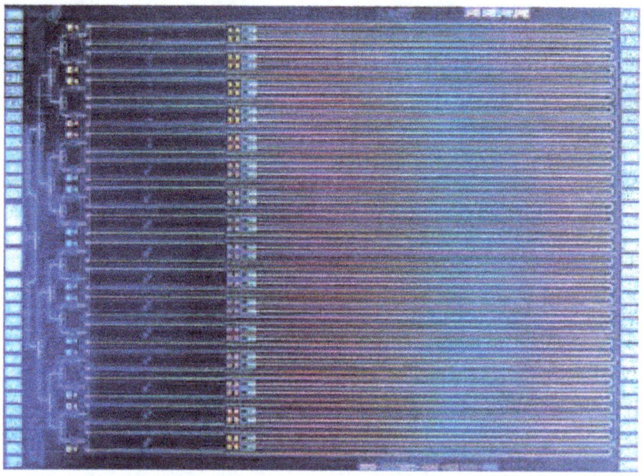

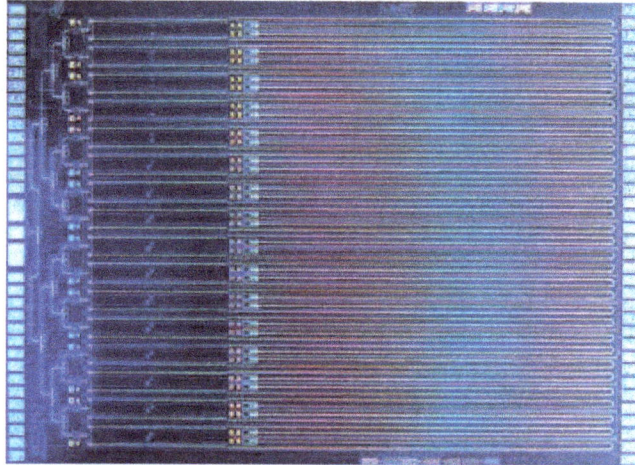

Image of a Josephson voltage standard chip containing approximately 300,000 microfabricated Josephson junctions.

The superconducting tunnel junction (STJ) — also known as a superconductor–insulator–superconductor tunnel junction (SIS) — is an electronic device consisting of two superconductors separated by a very thin layer of insulating material. Current passes through the junction via the process of quantum tunneling. The STJ is a type of Josephson junction, though not all the properties of the STJ are described by the Josephson effect.

These devices have a wide range of applications, including high-sensitivity detectors of electromagnetic radiation, magnetometers, high speed digital circuit elements, and quantum computing circuits.

### Quantum Tunneling

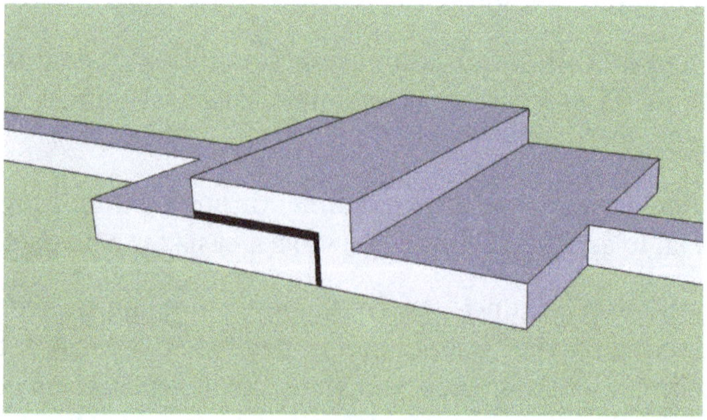

Illustration of a thin-film superconducting tunnel junction (STJ). The superconducting material is light blue, the insulating tunnel barrier is black, and the substrate is green.

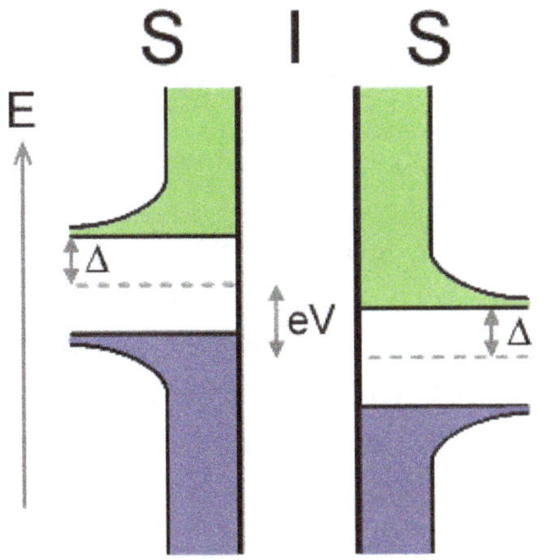


Energy diagram of a superconducting tunnel junction. The vertical axis is energy, and the horizontal axis shows the density of states. Cooper pairs exist at the Fermi energy, indicated by the dashed lines. A bias voltage V is applied across the junction, shifting the Fermi energies of the two superconductors relative to each other by an energy eV, where e is the electron charge. Quasiparticle states exist for energies greater than Δ from the Fermi energy, where Δ is the superconducting energy gap. Green and blue indicate empty and filled quasiparticle states, respectively, at zero temperature.

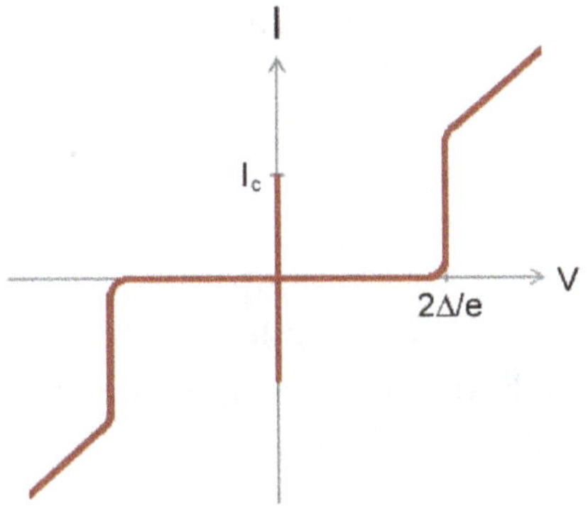


Sketch of the current-voltage (I-V) curve of a superconducting tunnel junction. The Cooper pair tunneling current is seen at V = 0, while the quasiparticle tunneling current is seen for V > 2Δ/e and V < -2Δ/e.

All currents flowing through the STJ pass through the insulating layer via the process of quantum tunneling. There are two components to the tunneling current. The first is from the tunneling of Cooper pairs. This supercurrent is described by the ac and dc Josephson relations, first predicted by Brian David Josephson in 1962. For this prediction, Josephson received the Nobel prize in physics in 1973. The second is the quasiparticle current, which, in the limit of zero temperature, arises when the energy from the bias voltage $eV$ exceeds twice the value of superconducting energy gap Δ. At finite temperature, a small quasiparticle tunneling current – called the subgap current – is present even for voltages less than twice the energy gap due to the thermal promotion of quasiparticles above the gap.

If the STJ is irradiated with photons of frequency $f$, the dc current-voltage curve will exhibit both Shapiro steps and steps due to photon-assisted tunneling. Shapiro steps arise from the response of the supercurrent and occur at voltages equal to $nhf/(2e)$, where $h$ is Planck's constant, $e$ is the electron charge, and $n$ is an integer. Photon-assisted tunneling arises from the response of the quasiparticles and gives rise to steps displaced in voltage by $nhf / e$ relative to the gap voltage.

## Device Fabrication

The device is typically fabricated by first depositing a thin film of a superconducting metal such as aluminum on an insulating substrate such as silicon. The deposition is performed inside a vacuum chamber. Oxygen gas is then introduced into the chamber, resulting in the formation of an insulating layer of aluminum oxide ( $Al_2O_3$ ) with a typical thickness of several nanometers. After the vacuum is restored, an overlapping layer of superconducting metal is deposited, completing the STJ. To create a well-defined overlap region, a procedure known as the Niemeyer-Dolan technique is commonly used. This technique uses a suspended bridge of resist with a double-angle deposition to define the junction.

Aluminum is widely used for making superconducting tunnel junctions because of its unique ability to form a very thin (2-3 nm) insulating oxide layer with no defects that short-circuit the insulating layer. The superconducting critical temperature of aluminum is approximately 1.2 kelvin (K). For many applications, it is convenient to have a device that is superconducting at a higher temperature, in particular at a temperature above the boiling point of liquid helium, which is 4.2 K at atmospheric pressure. One approach to achieving this is to use niobium, which has a superconducting critical temperature in bulk form of 9.3 K. Niobium, however, does not form an oxide that is suitable for making tunnel junctions. To form an insulating oxide, the first layer of niobium can be coated with a very thin layer (approximately 5 nm) of aluminum, which is then oxidized to form a high quality aluminum oxide tunnel barrier before the final layer of niobium is deposited. The thin aluminum layer is proximitized by the thicker niobium, and the resulting device has a superconducting critical temperature above 4.2 K. Early work used lead-lead oxide-lead tunnel junctions. Lead has a superconducting critical temperature of 7.2 K in bulk form, but lead oxide tends to develop defects (sometimes called pinhole defects) that short-circuit the tunnel barrier when the device is thermally cycled between cryogenic temperatures and room temperature, and as result lead is no longer widely used to make STJs.

## Applications

### Radio Astronomy

STJs are the most sensitive heterodyne receivers in the 100 GHz to 1000 GHz frequency range, and hence are used for radio astronomy at these frequencies. In this application, the STJ is dc biased at a voltage just below the gap voltage ($|V| = 2\Delta/e$). A high frequency signal from an astronomical object of interest is focused onto the STJ, along with a local oscillator source. Photons absorbed by the STJ allow quasiparticles to tunnel via the process of photon-assisted tunneling. This photon-assisted tunneling changes the current-voltage curve, creating a nonlinearity that produces an output at the difference frequency of the astronomical signal and the local oscillator. This output is a frequency down-converted version of the astronomical signal. These receivers are so sensitive that an accurate description of the device performance must take into account the effects of quantum noise.

### Single-photon Detection

In addition to heterodyne detection, STJs can also be used as direct detectors. In this application, the STJ is biased with a dc voltage less than the gap voltage. A photon absorbed in the superconductor breaks Cooper pairs and creates quasiparticles. The quasiparticles tunnel across the junction in the direction of the applied voltage, and the resulting tunneling current is proportional to the photon energy. STJ devices have been employed as single-photon detectors for photon frequencies ranging from X-rays to the infrared.

## SQUIDs

The superconducting quantum interference device or SQUID is based on a superconducting loop containing Josephson junctions. SQUIDs are the world's most sensitive magnetometers, capable of measuring a single magnetic flux quantum.

## Quantum Computing

Superconducting quantum computing utilizes STJ-based circuits, including charge qubits, flux qubits and phase qubits.

## RSFQ

The STJ is the primary active element in rapid single flux quantum or RSFQ fast logic circuits.

## Josephson Voltage Standard

When a high frequency current is applied to a Josephson junction, the ac Josephson current will synchronize with the applied frequency giving rise to regions of constant voltage in the I-V curve of the device (Shapiro steps). For the purpose of voltage standards, these steps occur at the voltages $nf / K_J$ where $n$ is an integer, $f$ is the applied frequency and the Josephson constant $K_J = 483597.9\text{GHz} / \text{V}$ is an internationally defined constant essentially equal to $2e / h$. These steps provide an exact conversion from frequency to voltage. Because frequency can be measured with very high precision, this effect is used as the basis of the Josephson voltage standard, which implements the international definition of the " conventional" volt.

## Superconducting Energy Gap

Using Bogoliubov-Valatin canonical transformation we have rewritten the pair Hamiltonian as

$$H_m = \sum_{\vec{k}} \xi_{\vec{k}} \left\{ \left( |u_{\vec{k}}|^2 - |v_{\vec{k}}|^2 \right) \left( \gamma_{\vec{k}0}^+ \gamma_{\vec{k}0} + \gamma_{\vec{k}1}^+ \gamma_{\vec{k}1} \right) + 2u_{\vec{k}}^* v_{\vec{k}}^* \gamma_{\vec{k}1} \gamma_{\vec{k}0} + 2u_{\vec{k}} v_{\vec{k}} \gamma_{\vec{k}0}^+ \gamma_{\vec{k}1}^+ + 2 |v_{\vec{k}}|^2 \right\}$$

$$+ \sum_{\vec{k}} \left\{ \left( u_{\vec{k}} v_{\vec{k}}^* - u_{\vec{k}}^* v_{\vec{k}} \right) \left( \gamma_{\vec{k}0}^+ \gamma_{\vec{k}0} + \gamma_{\vec{k}1}^+ \gamma_{\vec{k}1} \right) + \left( v_{\vec{k}}^2 - u_{\vec{k}}^2 \right) \left( \gamma_{\vec{k}1} \gamma_{\vec{k}0} + \gamma_{\vec{k}0}^+ \gamma_{\vec{k}1}^+ \right) + b_{\vec{k}} \right\}$$

In order to make the Hamiltonian diagonal, we put the coefficients of $\gamma_{\vec{k}1}\gamma_{\vec{k}0}$ and $\gamma_{\vec{k}0}^+\gamma_{\vec{k}1}^+$ equal to zero which results in the equation

$$2\xi_{\vec{k}} u_{\vec{k}} v_{\vec{k}} + \Delta_{\vec{k}}^* u_{\vec{k}}^2 = 0$$

multiplying by $\frac{\Delta_{\vec{k}}^*}{u_{\vec{k}}^2}$

$$\xi_{\vec{k}} \frac{\Delta_{\vec{k}}^*}{u_{\vec{k}}^2} + \Delta_{\vec{k}}^{*2} \frac{v_{\vec{k}}^2}{u_{\vec{k}}^2} - \Delta_{\vec{k}} \Delta_{\vec{k}}^* = 0$$

Or,

$$\Delta_{\vec{k}}^{*2} \frac{v_{\vec{k}}^2}{u_{\vec{k}}^2} + 2\xi_{\vec{k}} \frac{\Delta_{\vec{k}}^* v_{\vec{k}}}{u_{\vec{k}}} - \left|\Delta_{\vec{k}}\right|^2 = 0$$

Using $\frac{v_{\vec{k}}}{u_{\vec{k}}}$ as x,we have using the quadratic formula with $\frac{v_{\vec{k}}}{u_{\vec{k}}}$ as the variable,

$$\frac{v_{\vec{k}}}{u_{\vec{k}}} = \frac{-2\xi_{\vec{k}}\Delta_{\vec{k}}^* \pm \sqrt{\left(2\xi_{\vec{k}}\Delta_{\vec{k}}^*\right)^2 + 4\Delta_{\vec{k}}^{*2}\left|\Delta_{\vec{k}}\right|^2}}{2\Delta_{\vec{k}}^{*2}}$$

$$= \frac{-\xi_{\vec{k}}}{\Delta_{\vec{k}}^*} + \frac{2\Delta_{\vec{k}}^*}{2\Delta_{\vec{k}}^{*2}} \sqrt{\xi_{\vec{k}}^2 + \left|\Delta_{\vec{k}}\right|^2}$$

Or,

$$\frac{\Delta_{\vec{k}}^* v_{\vec{k}}}{u_{\vec{k}}} = \sqrt{\xi_{\vec{k}}^2 + \left|\Delta_{\vec{k}}\right|^2} - \xi_{\vec{k}}$$

We would have gotten the same result if we had treated $\Delta_{\vec{k}}^* \frac{v_{\vec{k}}}{u_{\vec{k}}}$ as the variable. Now

$$\Delta_{\vec{k}}^* \frac{v_{\vec{k}}}{u_{\vec{k}}} = E_{\vec{k}} - \xi_{\vec{k}}$$

From which we get

$$\left|\frac{v_{\vec{k}}}{u_{\vec{k}}}\right| = \frac{E_{\vec{k}} - \xi_{\vec{k}}}{\left|\Delta_{\vec{k}}\right|}$$

We also have the normalization condition

$$\left|u_{\vec{k}}\right|^2 + \left|v_{\vec{k}}\right|^2 = 1$$

so that

$$\left|v_{\vec{k}}\right|^2 = 1 - \left|u_{\vec{k}}\right|^2$$

$$= \frac{1}{2}\left[1 - \frac{\xi_{\vec{k}}}{E_{\vec{k}}}\right]$$

and

$$\left|u_{\vec{k}}\right|^2 = \frac{1}{2}\left[1 + \frac{\xi_{\vec{k}}}{E_{\vec{k}}}\right]$$

which gives the values of the coefficients $u_{\vec{k}}$ and $v_{\vec{k}}$ used in defining the Bogoliubov-Valatin transformation for diagonalizing the model Hamiltonian $H_m$.

Next we write down terms that are needed for the Hamiltonian,

$$\Delta_{\vec{k}}^* \frac{v_{\vec{k}}}{u_{\vec{k}}} = \left(E_{\vec{k}} - \xi_{\vec{k}}\right) = \left(\xi_{\vec{k}}^2 + \left|\Delta_{\vec{k}}\right|^2\right)^{\frac{1}{2}} - \xi_{\vec{k}}$$

$$\Delta_{\vec{k}}^* v_{\vec{k}} u_{\vec{k}}^* = \Delta_{\vec{k}}^* \frac{v_{\vec{k}}}{u_{\vec{k}}} u_{\vec{k}}^* \; u_{\vec{k}}$$

$$= \Delta_{\vec{k}}^* \frac{v_{\vec{k}}}{u_{\vec{k}}} \left|u_{\vec{k}}\right|^2$$

$$= \left(E_{\vec{k}} - \xi_{\vec{k}}\right) \left|u_{\vec{k}}\right|^2$$

$$= \left(E_{\vec{k}} - \xi_{\vec{k}}\right) \frac{\left(E_{\vec{k}} + \xi_{\vec{k}}\right)}{2E_{\vec{k}}}$$

Or,

$$\Delta_{\vec{k}}^* v_{\vec{k}} u_{\vec{k}}^* = \frac{E_{\vec{k}}^2 - \xi_{\vec{k}}^2}{2E_{\vec{k}}}$$

Thus,

$$\Delta_{\vec{k}}^* v_{\vec{k}} u_{\vec{k}}^* = \frac{E_{\vec{k}}^2 - \xi_{\vec{k}}^2}{2E_{\vec{k}}}$$

$$\left|u_{\vec{k}}\right|^2 - \left|v_{\vec{k}}\right|^2 = \frac{1}{2}\left(1 + \frac{\xi_{\vec{k}}}{E_{\vec{k}}}\right) - \frac{1}{2}\left(1 - \frac{\xi_{\vec{k}}}{E_{\vec{k}}}\right)$$

$$= \frac{\xi_{\vec{k}}}{E_{\vec{k}}}$$

The diagonalized Hamiltonian is

$$H_m = \sum_{\vec{k}} \xi_{\vec{k}} \left[ \left( \left| u_{\vec{k}} \right|^2 - \left| v_{\vec{k}} \right|^2 \right) \left( \gamma_{\vec{k}0}^+ \gamma_{\vec{k}0} + \gamma_{\vec{k}1}^+ \gamma_{\vec{k}1} \right) + 2 \left| v_{\vec{k}} \right|^2 \right]$$

$$+ \sum_{\vec{k}} \left[ \left( \Delta_{\vec{k}} u_{\vec{k}} v_{\vec{k}}^* + \Delta_{\vec{k}}^* u_{\vec{k}}^* v_{\vec{k}} \right) \left( \gamma_{\vec{k}0}^+ \gamma_{\vec{k}0} + \gamma_{\vec{k}1}^+ \gamma_{\vec{k}1} - 1 \right) + \Delta_{\vec{k}} b_{\vec{k}}^* \right]$$

$$= \sum_{\vec{k}} \xi_{\vec{k}} \left[ \frac{\xi_{\vec{k}}}{E_{\vec{k}}} \left( \gamma_{\vec{k}0}^+ \gamma_{\vec{k}0} + \gamma_{\vec{k}1}^+ \gamma_{\vec{k}1} \right) + \left( \frac{E_{\vec{k}} - \xi_{\vec{k}}}{E_{\vec{k}}} \right) \right]$$

$$+ \sum_{\vec{k}} \left[ \left( \frac{E_{\vec{k}}^2 - \xi_{\vec{k}}^2}{2E_{\vec{k}}} + \frac{E_{\vec{k}}^2 - \xi_{\vec{k}}^2}{2E_{\vec{k}}} \right) \left( \gamma_{\vec{k}0}^+ \gamma_{\vec{k}0} + \gamma_{\vec{k}1}^+ \gamma_{\vec{k}1} - 1 \right) + \Delta_{\vec{k}} b_{\vec{k}}^* \right]$$

$$= \sum \frac{\xi_{\vec{k}}}{E_{\vec{k}}} \left( \gamma_{\vec{k}0}^+ \gamma_{\vec{k}0} + \gamma_{\vec{k}1}^+ \gamma_{\vec{k}1} \right) + \frac{\xi_{\vec{k}} E_{\vec{k}} - \xi_{\vec{k}}^2}{E_{\vec{k}}}$$

$$+ \sum_{\vec{k}} \frac{E_{\vec{k}}^2 - \xi_{\vec{k}}^2}{E_{\vec{k}}} \left( \gamma_{\vec{k}0}^+ \gamma_{\vec{k}0} + \gamma_{\vec{k}1}^+ \gamma_{\vec{k}1} \right) - \frac{E_{\vec{k}}^2 - \xi_{\vec{k}}^2}{E_{\vec{k}}} + \Delta_{\vec{k}} b_{\vec{k}}^*$$

Finally, the model Hamiltonian, in the diagonal form, becomes

$$H_m = \sum_{\vec{k}} \xi_{\vec{k}} - E_{\vec{k}} + \Delta_{\vec{k}} b_{\vec{k}}^* + \sum_{\vec{k}} E_{\vec{k}} \left( \gamma_{\vec{k}0}^+ \gamma_{\vec{k}0} + \gamma_{\vec{k}1}^+ \gamma_{\vec{k}1} \right)$$

The first term gives the condensation energy while the second term is the energy associated with quasiparticle excitations. Now we look at the superconducting energy gap.

We have defined the energy gap parameter $\Delta_{\vec{k}}$ as

$$\Delta_{\vec{k}} = -\sum_l V_{\vec{k}l} b_{\vec{l}}$$

$$= -\sum_{\vec{l}} V_{\vec{k}\vec{l}} b_{\vec{l}} < C_{-\vec{l}\downarrow} C_{\vec{l}\uparrow} >$$

We have also seen that (ignoring the off-diagonal terms since they do not contribute)

$$C_{-\vec{l}\downarrow} C_{\vec{l}\uparrow} = -v_{\vec{k}} u_{\vec{k}}^* \gamma_{\vec{k}0}^+ \gamma_{\vec{k}0} + u_{\vec{k}}^* v_{\vec{k}} \left( 1 - \gamma_{\vec{k}1}^+ \gamma_{\vec{k}} \right)$$

$$= u_{\vec{k}}^* v_{\vec{k}} \left( 1 - \gamma_{\vec{k}0}^+ \gamma_{\vec{k}0} - \gamma_{\vec{k}1}^+ \gamma_{\vec{k}1} \right)$$

Therefore, we can write the gap parameter as,

$$\Delta_{\vec{k}} = -\sum_{\vec{l}} V_{\vec{k}\vec{l}} u_{\vec{k}}^* v_{\vec{l}} < 1 - \gamma_{\vec{k}0}^+ \gamma_{\vec{k}0} - \gamma_{\vec{k}1}^+ \gamma_{\vec{k}1}$$

we note that $\gamma_{\vec{k}0}^+ \gamma_{\vec{k}0}$ and $\gamma_{\vec{k}1}^+ \gamma_{\vec{k}1}$ are number operators for quasiparticles. The excitations are given by the Fermi-Dirac distribution function

$$f\left(E_{\vec{k}}\right) = \frac{1}{e^{E_{\vec{k}}/kT} + 1}$$

and at $T = 0$, we have $f\left(E_{\vec{k}}\right) \approx 0$, therefore

$$\Delta_{\vec{k}} = -\sum_{\vec{l}} V_{\vec{k}l} u_{\vec{l}}^* v_{\vec{l}}$$

we have seen earlier that

$$\Delta_{\vec{l}} v_{\vec{l}}^* u_{\vec{l}} = \frac{E_{\vec{l}}^2 - \xi_{\vec{l}}^2}{2E_{\vec{l}}}$$

$$\Rightarrow v_{\vec{l}}^* u_{\vec{l}} = \frac{\left(\Delta_{\vec{l}}^2 + \xi_{\vec{l}}^2\right) - \xi_{\vec{l}}^2}{2E_{\vec{l}} \Delta_{\vec{l}}}$$

Or,

$$v_{\vec{l}}^* u_{\vec{l}} = \frac{\Delta_{\vec{l}}}{2E_{\vec{l}}}$$

Substituting in the gap equation, Equation, we get

$$\Delta_{\vec{k}} = -\sum_{\vec{l}} V_{\vec{k}l} \frac{\Delta_{\vec{l}}}{2E_{\vec{l}}}$$

Assuming a constant interaction potential $V_{\vec{k}l} = V$

$$\Delta_{\vec{k}} = -V \sum_{\vec{l}} \frac{\Delta_{\vec{l}}}{2E_{\vec{l}}}$$

Next, let us assume that the gap parameter is isotropic,

$$\Delta = -V\Delta \sum_{\vec{l}} \frac{1}{2E_{\vec{l}}}$$

Or,

$$\frac{1}{V} = -\sum_{\vec{l}} \frac{1}{2E_{\vec{l}}}$$

Or,

$$\frac{1}{V} = N(0)\int_0^{\hbar\omega_D} \frac{d\xi}{\sqrt{\Delta^2 + \xi^2}} = \sinh^{-1} \frac{\hbar\omega_D}{\Delta}$$

$$= \sinh^{-1}\frac{\hbar\omega_D}{\Delta}$$

$$\Delta(0) = \frac{\hbar\omega_d}{\sinh\left[1/VN(0)\right]}$$

$$\simeq 2\hbar\omega_d e^{-1/(VN(0))}$$

Thus, the gap parameter at $T = 0$, is approximately given by

$$\Delta(0) \simeq 2\hbar\omega_d e^{-1/(VN(0))}$$

## Superconducting Transition Temperature

At finite temperatures, the Fermi-Dirac function can be written as

$$f\left(E_{\vec{k}}\right) = \frac{1}{e^{E_{\vec{k}}/k_{\vec{l}1}T} + 1}$$

so that

$$< 1 - \gamma_{\vec{l}0}^{+}\gamma_{\vec{l}0} - \gamma_{\vec{l}1}^{+}\gamma_{\vec{l}1} > = 1 - 2f\left(E_{\vec{k}}\right)$$

$$= 1 - \frac{2}{e^{\beta E_{\vec{l}}} + 1}$$

$$= \frac{e^{\beta E_{\vec{l}}} - 1}{e^{\beta E_{\vec{l}}} + 1}$$

$$= \tanh\frac{\beta E_{\vec{l}}}{2}$$

where $\beta = \frac{1}{\vec{k}T}$. We have,

$$< 1 - \gamma_{\vec{l}0}^{+}\gamma_{\vec{l}0} - \gamma_{\vec{l}1}^{+}\gamma_{\vec{l}1} > = \tanh\frac{\beta E_{\vec{l}}}{2}$$

Thus , the gap equation becomes

$$\Delta_{\vec{k}} = -\sum_{\vec{l}} V_{\vec{k}\vec{l}} u_{\vec{l}}^{*} v_{\vec{l}} < 1 - 2f\left(E_{\vec{l}}\right) >$$

$$= \sum_{\vec{l}} V_{\vec{k}\vec{l}} \frac{\Delta_{\vec{l}}}{2E_{\vec{l}}} \tanh\frac{E_{\vec{l}}}{2k_B T}$$

$$= \sum_{\vec{l}} V_{\vec{k}\vec{l}} \frac{\Delta_{\vec{l}}}{2E_{\vec{l}}} \tanh\frac{E_{\vec{l}}}{2k_B T}$$

In the BCS approximation,

$$V_{\vec{k}\vec{l}} = \begin{cases} -V & E_{\vec{k}} < \varepsilon_F < \hbar\omega_D \\ ::0 & otherwise \end{cases}$$

and

$$\Delta_{\vec{l}} = \Delta_{\vec{k}} = \Delta$$

so that

$$\frac{1}{V} = \frac{1}{2}\sum_{\vec{k}} \frac{\tanh \frac{E_{\vec{k}}}{2k_B T}}{E_{\vec{k}}}$$

Changing summation into integration and the change of variable $\xi = \frac{2x}{\beta_c}$, at $T = 0$,

$$\frac{1}{VN(0)} = \int_0^{\frac{\beta_c \hbar\omega_D}{2}} \frac{\tanh x}{x} dx = 1n\left(\frac{A\beta_c \hbar\omega_D}{2}\right)$$

$$= 1n\left(\frac{A\beta_c \hbar\omega_D}{2}\right)$$

Or,

$$e^{1/(VN(0)) = \frac{A\beta_c \hbar\omega_D}{2}]}$$

In the above equation, $A$ is given by

$$A = 2e^{\gamma/\pi}$$

$$\approx 1.13$$

where $\gamma = 0.577...$, and it is known as the Euler's constant. Thus,

$$\beta = \frac{2e^{1/(VN(0))}}{A\hbar\omega_D}$$

The superconducting transition temperature, in the BCS approximation, is found to be

$$k_B T_c = A\hbar\omega_D e^{-1/(VN(0))}$$

$$= 1.13\hbar\omega_D e^{-1/(VN(0))}$$

A comparison with the equation for the gap parameter yields

$$\frac{\Delta(0)}{\vec{k}_B T_c} = \frac{2}{1.13} = 1.764$$

## Temperature Dependence of the Superconducting Gap

We have seen earlier that

$$\frac{1}{V}=\frac{1}{2}\sum_{\vec{k}}\frac{\tanh\frac{E_{\vec{k}}}{2k_BT}}{E_{\vec{k}}}$$

Changing to integration, as before, we have

$$\Rightarrow\frac{1}{VN(0)}=\int_0^{\hbar\omega_D}\frac{\tanh\frac{1}{2}\beta\left(\xi^2+\Delta^2\right)^{\frac{1}{2}}}{\left(\xi^2+\Delta^2\right)^{\frac{1}{2}}}d\xi$$

for $\frac{\hbar\omega_D}{k_BT}>>1,$ we find that close to $T_c$,

$$\frac{\Delta(T)}{\Delta(0)}\simeq 1.74\left(1-\frac{T}{T_c}\right)^{\frac{1}{2}}$$

The temperature dependence of the gap parameter, also known as the order parameter, is shown in Figure.

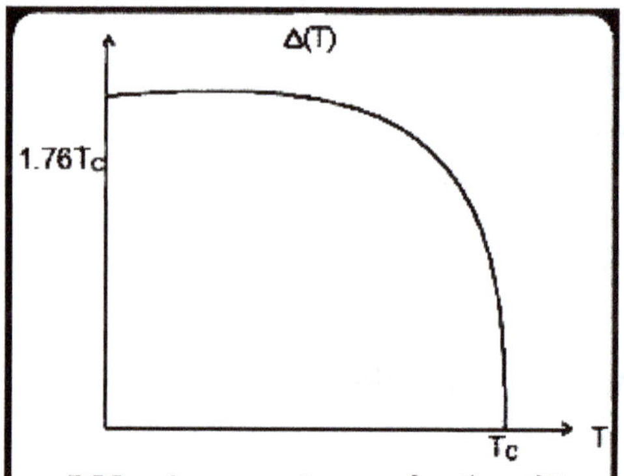

BCS order parameter as a function of T

## Coherence Length and the Pair Wavefunction

In the superconducting state one can define a characteristic length, called coherence length, $\zeta(T)$, as

$$\zeta(T)=\frac{\hbar v_F}{\pi\Delta(T)}$$

where $v_F$ is the Fermi velocity. The value of the coherence length at $T=0$, known as $\zeta_0$, is of the order of $10^{-4}$ cm. The coherence length is much larger than the separation between two electrons, leading to overlap of many pair wavefunctions. The pair wavefunction is given by

$$\psi_{\vec{k}}=-u_{\vec{k}}v_{\vec{k}}$$

so that in the real space the wavefunction can be written as

$$\Psi(\vec{r}) = \sum_{\vec{k}} \psi_{\vec{k}} e^{i\vec{k}\cdot\vec{r}}$$
$$= -\sum_{\vec{k}} u_{\vec{k}} v_{\vec{k}} e^{i\vec{k}\cdot\vec{r}}$$
$$= \frac{1}{2}\sum_{\vec{k}} \left( \frac{|\Delta|^2}{\xi_{\vec{k}^2} + |\Delta|^2} \right)^{1/2} e^{i\vec{k}\cdot\vec{r}}$$

## Charge Density

The charge density of the BCS condensate can be written as

$$\rho = e\sum_{\vec{k}\sigma} < C^{+}_{\vec{k}\sigma} C_{\vec{k}\sigma} >$$
$$= Q + Q*$$

where

$$Q = 2\sum_{\vec{k}} |v_{\vec{k}}|^2$$

and

$$Q* = 2\sum \left( |u_{\vec{k}}|^2 - |v_{\vec{k}}|^2 \right) f\left(E_{\vec{k}}\right)$$

The first term $Q$ is the contribution from the electron pairs while the second term $Q*$ comes from the quasiparticle excitations.

## Heat Capacity and other Thermodynamic Properties

In the following we like to discuss some of the properties of the superconducting state. These properties change substantially from their normal state values. The first thing that we look at is the change in the heat capacity as the metal goes from the normal state to the superconducting state.

The electronic entropy in the superconducting state can be written as

$$S_{es} = 2\vec{k}_{\beta} \sum_{\vec{k}} \left[ \left(1 - f_{\vec{k}}\right) ln\left(1 - f_{\vec{k}}\right) + f_{\vec{k}} ln\, f_{\vec{k}} \right]$$

from which the heat capacity, $C_{es,}$ can be found as

$$C_{es} = T\frac{dS_{es}}{dT}$$

$$= -\beta \frac{dS_{es}}{d\beta}$$

where $\beta = \frac{1}{\vec{k}T}$.

Since, $f_{\vec{k}} = 1/\left(e^{\beta E_{\vec{k}}} + 1\right)$, we have

$$C_{es} = +\beta 2k_B \sum_{\vec{k}} -\frac{1}{1-f_{\vec{k}}}\frac{\partial f_{\vec{k}}}{\partial \beta} - \frac{\partial f_{\vec{k}}}{\partial \beta} ln\left(1-f_{\vec{k}}\right) + \frac{f_{\vec{k}}}{1-f_{\vec{k}}}\frac{\partial f_{\vec{k}}}{\partial \beta} + \frac{\partial f_{\vec{k}}}{\partial \beta} lnf_{\vec{k}} + \frac{f_{\vec{k}}}{f_{\vec{k}}}\frac{\partial f_{\vec{k}}}{\partial \beta}$$

$$= 2\beta k_B \sum_{\vec{k}} \frac{\partial f_{\vec{k}}}{\partial \beta}\left[-\frac{1}{1-f_{\vec{k}}} - ln\left(1-f_{\vec{k}}\right) + \frac{f_{\vec{k}}}{1-f_{\vec{k}}} + lnf_{\vec{k}} + 1\right]$$

$$= 2\beta k_B \sum_{\vec{k}} \frac{\partial f_{\vec{k}}}{\partial \beta}\left[\frac{-1-\left(1-f_{\vec{k}}\right)ln\left(1-f_{\vec{k}}\right) + f_{\vec{k}} + lnf_{\vec{k}}\left(1-f_{\vec{k}}\right) + \left(1-f_{\vec{k}}\right)}{\left(1-f_{\vec{k}}\right)}\right]$$

$$= 2\beta k_B \sum_{\vec{k}} \frac{\partial f_{\vec{k}}}{\partial \beta}\left[\frac{-ln\left(1-f_{\vec{k}}\right) + lnf_{\vec{k}}\left(1-f_{\vec{k}}\right) + \left(1-f_{\vec{k}}\right)lnf_{\vec{k}}}{\left(1-f_{\vec{k}}\right)}\right]$$

$$= 2\beta k_B \sum_{\vec{k}} \frac{\partial f_{\vec{k}}}{\partial \beta}\left[\frac{-ln\left(1-f_{\vec{k}}\right)\left(1-f_{\vec{k}}\right) + 1\left(1-f_{\vec{k}}\right)lnf_{\vec{k}}}{\left(1-f_{\vec{k}}\right)}\right]$$

$$= 2\beta k_B \sum_{\vec{k}} \frac{\partial f_{\vec{k}}}{\partial \beta}\left[lnf_{\vec{k}} - ln\left(1-f_{\vec{k}}\right)\right]$$

$$= 2\beta k_B \sum_{\vec{k}} \frac{\partial f_{\vec{k}}}{\partial \beta}\left[1n\frac{f_{\vec{k}}}{1-f_{\vec{k}}}\right]$$

To further simplify the expression for the heat capacity we need to consider the Fermi-Dirac function. Using

$$f_{\vec{k}} = \frac{1}{e^{\beta E_{\vec{k}}} + 1}; \qquad 1 - f_{\vec{k}} = 1 - \frac{1}{e^{\beta E_{\vec{k}}} + 1} = \frac{e^{\beta E_{\vec{k}}} + 1 - 1}{e^{\beta E_{\vec{k}}} + 1}$$

$$1 - f_{\vec{k}} = 1 - \frac{1}{e^{\beta E_{\vec{k}}} + 1}$$
$$= \frac{e^{\beta E_{\vec{k}}}}{e^{\beta E_{\vec{k}}} + 1}$$

Thus,

$$\frac{f_{\vec{k}}}{1-f_{\vec{k}}} = \frac{1}{e^{\beta E_{\vec{k}}} + 1} \times \frac{e^{\beta E_{\vec{k}}} + 1}{e^{\beta E_{\vec{k}}}}$$
$$= e^{-\beta E_{\vec{k}}}$$

Therefore,

$$ln\frac{f_{\vec{k}}}{1-f_{\vec{k}}}=\left(-\beta E_{\vec{k}}\right)$$

Hence the heat capacity becomes

$$C_{es}=2\beta k_B\sum_{\vec{k}}\frac{\partial f_{\vec{k}}}{\partial\beta}\left\{-\beta E_{\vec{k}}\right\}$$

$$=2\beta^2 k_B\sum_{\vec{k}}E_{\vec{k}}\frac{\partial f_{\vec{k}}}{\partial\beta}$$

We can rewrite the right hand side by noting that

$$\frac{\partial f_{\vec{k}}}{\partial\beta}=\frac{df_{\vec{k}}}{d\left(\beta E_{\vec{k}}\right)}\frac{d\left(\beta E_{\vec{k}}\right)}{d\beta}=\frac{df_{\vec{k}}}{d\left(\beta E_{\vec{k}}\right)}\left[E_{\vec{k}}+\beta\frac{dE_{\vec{k}}}{d\beta}\right]$$

Substituting it in the expression for $C_{es}$, we have

$$C_{es}=-2\beta^2\,\mathrm{k}_B\left[\frac{df_{\vec{k}}}{d\left(\beta E_{\vec{k}}\right)}E_{\vec{k}}^2+\frac{1}{2}E_{\vec{k}}\frac{dE_{\vec{k}}}{d\beta}\right]$$

Or,

$$C_{es}=2\beta\,\mathrm{k}_B\sum_{\vec{k}}\left[-\frac{\partial f_{\vec{k}}}{\partial E_{\vec{k}}}\left(E_{\vec{k}}^2+\frac{1}{2}\beta\frac{d\Delta^2}{d\beta}\right)\right]$$

The first term in $C_{es}$ arises due to the rearrangement of the quasiparticles in the energy state due to temperature. The second term affects the occupation of quasiparticles through the temperature dependence of the superconducting gap.

For normal state $\Delta(T)\rightarrow 0$, and we see that the heat capacity can be written as

$$C_{en}=2\beta\,\mathrm{k}_B\sum_{\vec{k}}-\frac{\partial f_{\vec{k}}}{\partial E_{\vec{k}}}E_{\vec{k}}^2$$

Since $E_{\vec{k}}^2=\xi_{\vec{k}}^2$ in the normal state,

$$C_{en}=2\beta\,\mathrm{k}_B\sum_{\vec{k}}-\frac{\partial f_{\vec{k}}}{\partial\xi_{\vec{k}}}\xi_{\vec{k}}^2$$

$$=\frac{2\pi^2}{3}\mathrm{N}(0)\mathrm{k}_B^2\mathrm{T}$$

which is the heat capacity in the normal state.

The discontinuity in the heat capacity between the superconducting and the normal states can be determined, close to $T_c$, by changing the sum to an integral

$$\begin{aligned}\Delta C &= \left(C_{es} - C_{en}\right)\left|T_c\right. \\ &= N(0) k_B \beta^2 \left(\frac{d\Delta^2}{d\beta}\right)_{-\infty}^{\infty} \left(-\frac{\partial f}{\partial|\xi|}\right) d\xi \\ &= N(0)\left(-\frac{d\Delta^2}{dT}\right)\left|T - T_c\right.\end{aligned}$$

Near $T_c$ , we can write

$$\Delta^2 = \left[1.74\left(1.76 k_B T_c\right)\left(1 - \frac{T}{T_c}\right)\right]^2$$

so that

$$\Delta C = 9.4 N(0) k_B^2 T_c$$

At $T = T_c$, we see that the superconductivity leads to an increase in the heat capacity of the material. The normalized value of the discontinuity in the heat capacity is given by

$$\frac{\Delta C}{C_{en}} = \frac{9.4}{2\pi^2/3} = 1.43$$

Next we see how the heat capacity of the superconductor changes as $T \to 0$. As $T \to 0$, we find that $f_{\vec{k}} \sim e^{-\beta E_{\vec{k}}}$, and $\frac{d\Delta^2}{dT} \to 0$ so that the heat capacity can be written as

$$\begin{aligned}C_{es} &\approx 2\beta k_B \sum_{\vec{k}} \left[-\frac{\partial f_{\vec{k}}}{\partial E_{\vec{k}}}\left(E_{\vec{k}}^2\right)\right] \\ &= 2\beta^2 k_B \sum_{\vec{k}} \left[e^{-\beta E_{\vec{k}}}\left(\xi_{\vec{k}}^2 + \Delta_{\vec{k}}^2\right)\right]\end{aligned}$$

Changing the summation to an integration and noting that the integral contributes only when $\Delta >> \xi$, we have

$$\begin{aligned}C_{es} &\approx 2\beta^2 k_B N(0) \Delta_0^2 e^{-\beta\Delta_0}{}_0^{\infty} d\xi e^{-\beta\xi^2/(2\Delta_0)} \\ &= 2\beta^2 k_B N(0) \Delta_0^2 e^{-\beta\Delta_0} \frac{1}{2}\sqrt{\frac{\pi}{\beta/2\Delta_0}}\end{aligned}$$

where $\Delta_0 = \Delta(T = 0)$ is the maximum value of the gap parameter. The temperature variation of the heat capacity has an exponentially decaying term due to the superconducting energy gap.

# Quasiparticle Tunneling: Energy-Level Diagrams

For describing electron tunneling between metal-insulator-metal, metal-insulator-superconductor and superconductor-insulator-superconductor junctions we need to couple the junction through an interaction term in the Hamiltonian

$$H_T = \sum_{\sigma\vec{k}\vec{q}} T_{\vec{k}\vec{q}} C^{+}_{\vec{k}\sigma} C_{\vec{q}\sigma}$$

where $\vec{k}$ and $\vec{q}$ refer to the two metals on the two sides. It turns out that the coherence factors $u_{\vec{k}}$ and $v_{\vec{k}}$ drop out in a current calculations. Then it is easier if we assume a constant interaction matrix $T_{\vec{k}\vec{q}} = T$. Next we consider the three cases separately.

## Metal-Insulator-Metal Junction

Consider a metal-insulator-metal junction. In this case the current from metal 1 to metal 2 can be written as

$$I_{1-2} = A\int_{-\infty}^{\infty} |T|^2 N_1(E) f(E) N_2(E+eV)\left[1-f(E+eV)\right] dE$$

where $V$ is the applied voltage, $N_1(E)$ and $N_2(E)$ are the densities of states (DOS) on the two sides (normal or superconducting DOS as the case may be) and $f(E)$ and $f(E+eV)$ are Fermi-Dirac functions corresponding to energy $E$ and $E+eV$. Similarly, we get an expression for current from 2 to 1.

$$I_{2-1} = A\int_{-\infty}^{\infty} |T|^2 N_2(E+eV) f(E+eV) N_1(E)\left[1-f(E)\right]$$

The net current is the difference of these two tunneling currents, taking

$$N_1(E) \simeq N_1(0)$$
$$N_2(E+eV) \simeq N_2(0)$$

we can write the net current as

$$I_{nn} = A|T|^2 N_1(0) N_2(0) \int_{-\infty}^{\infty} \left[f(E) - f(E+eV)\right]$$

Since $eV$ is small, we can expand $f(E+eV)$ as

$$f(E) \simeq f(E) + \frac{\partial f}{\partial E}(eV)$$

Hence,

$$I_{nn} = A|T|^2 N_1(0) N_2(0) \int_{-\infty}^{\infty} (-eV) \frac{\partial f}{\partial E} dE$$

Using

$$-\frac{\partial f}{\partial E} = \delta(E)$$

we get

$$I_{nn} = A|T|^2 N_1(0) N_2(0) eV$$

Let $A|T|^2 N_1(0) N_2(0) e = G_{nn}$ normal conductance of the junction, then we can write the net current as

$$I_{nn} = G_{nn} V$$

## Metal-Insulator-Superconductor Junction

The net current in the case of a normal metal and a superconductor junction can be written as

$$I_{ns} = A|T|^2 \int_{-\infty}^{\infty} N_{1n}(E - eV) N_{2s}(E) [f(E - eV) - f(E)] dT$$

$$I_{ns} = A|T|^2 N_{1n}(0) N_{2n}(0) \int_{-\infty}^{\infty} [f(E - eV) - f(E)] dt$$

The current has to be evaluated numerically. However, one can look at the differential conductance defined as $G_{ns} = \frac{\partial I}{\partial V}$

$$\frac{\partial I}{\partial V} = \frac{G_{nn}}{e} \int_{-\infty}^{\infty} \frac{N_{2s}(E)}{N_{2n}(0)} \frac{\partial}{\partial V} [f(E - eV) - f(E)] dE$$

Using

$$f(E - eV) - f(E) = f(E) - eV \frac{\partial f}{\partial E}$$

we have for the differential conductance,

$$\frac{\partial I}{\partial V} = G_{nn} \int_{-\infty}^{\infty} \frac{N_{2s}(E)}{N_{2n}(0)} \frac{\partial f(E - eV)}{\partial(eV)} dE$$

$$G_{ns} = \frac{dI_{ns}}{dV}\bigg|_{T-0} = G_{nn}\frac{N_{2s}(e|V|)}{N_2(0)}$$

Thus, at low temperatures, the differential conductance measures the density of states.

## Superconductor-Insulator-Superconductor Junction

In the case of a superconductor-insulator-superconductor junction with superconducting gaps $\Delta_1$ and $\Delta_2$, the tunneling current is given by

$$I_{ss} = \frac{G_{nn}}{e}\int_{-\infty}^{\infty}\frac{N_{1s}(E-eV)}{N_1(0)}\frac{N_{2s}(E)}{N_2(0)}\left[f(E-eV)-f(E)\right]dE$$

$$I_{ss} = \frac{G_{nn}}{e}\int_{-\infty}^{\infty}\frac{|E-eV|}{\left((E-eV)^2+\Delta_1^2\right)^2}\frac{|E|}{\left[\left(E^2+\Delta_2^2\right)^{1/2}\right]}\left[f(E-eV)-f(E)\right]dE$$

Once again, the tunneling current can be calculated numerically. However, the qualitative features of the tunneling current in all the three cases are well described schematically as shown in Figures.

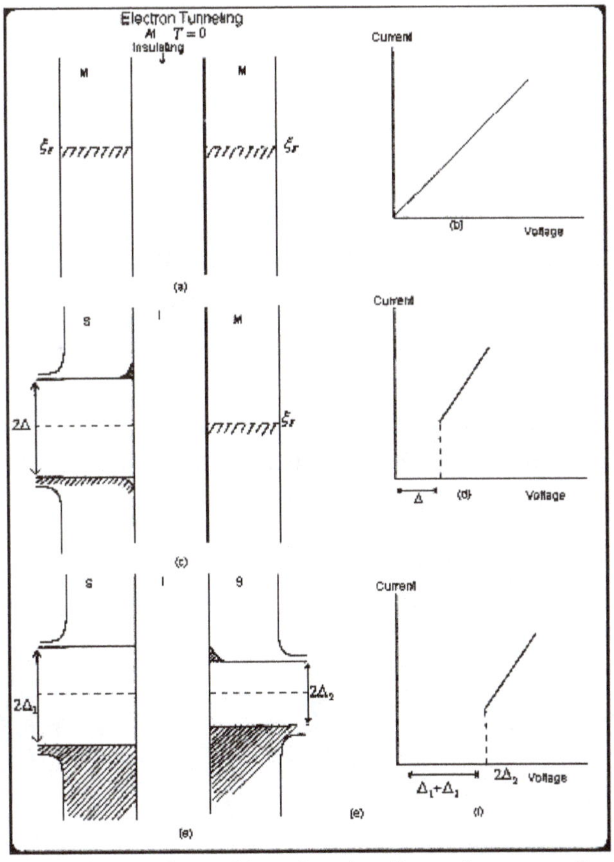


Figure shows the I - V curves for (a - b) metal-metal junction, (c - d) metal-superconductor junction and (e - f) superconductor-superconductor junction at T = 0. The electron occupied part is shown shaded.

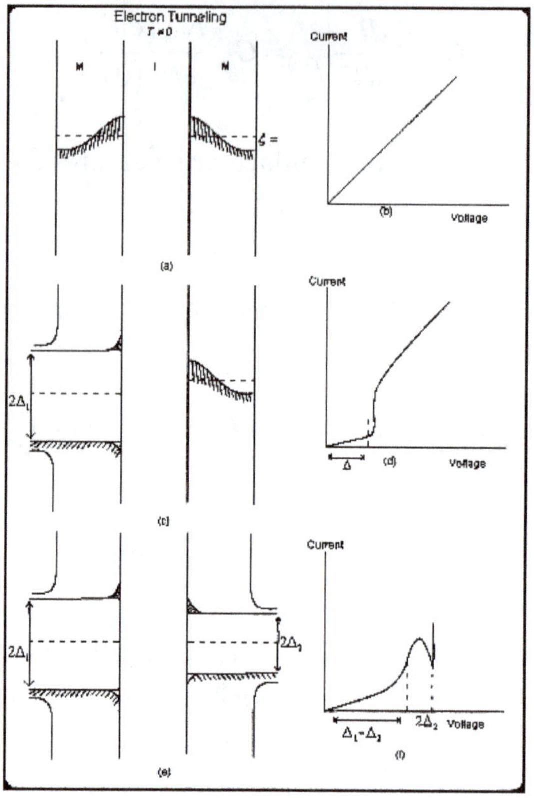


Figure shows the I - V curves for (a - b) metal-metal junction, (c - d) metal-superconductor junction and (e - f) superconductor-superconductor junction at T ≠ 0. The electron occupied part is shown shaded.

## Metal-Insulator-Metal Junction

For metal-insulator-metal junction the transfer Hamiltonian can be written as

$$H_t = \sum_{lr} T_{rl} \left[ c_r^+ c_l + hc \right]$$

where $l$ and $r$ indicate the materials on the left and right of the insulating material, respectively. Now, using the first-order, time-dependent perturbation theory, the charge crossing from right $(R)$ to left $(L)$ and from left-to-right can be calculated as

$$P_{L\to R} = 2\frac{2\pi}{\hbar}\sum_{lr} |T_{rl}|^2 (1-f_r) f_l \delta(\xi_l - eV - \xi_r)$$

and

$$P_{R\to L} = 2\frac{2\pi}{\hbar}\sum_{lr} |T_{rl}|^2 (1-f_l) f_r \delta(\xi_l - eV - \xi_r)$$

Note that while evaluating $\left[ < H_t^+ H_t > \right]$, we have

$$< c_n^+ c_{n'} > = f_n \delta_{nn'}$$

and

$$<c_n c_{n'}^+> = (1 - f_n)\delta_{nn'}$$

The total current across the insulating layer is given by

$$\begin{aligned} I &= e(P_{L\to R} - P_{R\to L}) \\ &= e2\frac{2\pi}{\hbar}\sum_{lr}|T_{rl}|^2 \delta(\xi_l - eV - \xi_r)(f_l - f_r f_l - f_r + f_r f_l) \\ &= \frac{4\pi e}{\hbar}\sum_{lr}|T_{rl}|^2 (f_l - f_r)\delta(\xi_l - eV - \xi_r) \end{aligned}$$

Summing over $l$ by replacing the summation by integration as

$$\sum_l \to \int N(\xi_l)d\xi_l,$$

where $N(\xi)$ is the density of states, we get

$$I = \frac{4\pi e}{\hbar}|T_{rl}|^2 N_l(\xi_l)\delta(\xi_l - eV - \xi_r)(f_e - f_r)d\xi_l$$

We can write the tunneling current as

$$I = \frac{4\pi e}{\hbar}|T|^2 \sum_r N_l(0)\int (f_e - f_r)\delta(\xi_l - eV - \xi_r)d\xi_l$$

where $|T|^2$ is the average of the matrix elements over the Fermi surface. Assuming $N_l(\xi_l)$ to be a constant over the range of integration, for the above integral we have

$$\int [f(\xi_l) - f(\xi_r)]\delta(\xi_l - eV - \xi_r)d\xi_l = f(eV + \xi_r) - f(\xi_r)$$

Since $eV$ is small we can write

$$\begin{aligned} f(eV + \xi_r) - f(\xi_r) &= f(\xi_r) + \frac{df}{d\varepsilon}\Big|_{\varepsilon_r} eV - f(\xi_r) \\ &= \delta(\xi_r)eV \end{aligned}$$

The expression for the net current reduces to,

$$\begin{aligned} I &= \frac{4\pi e}{\hbar}|T|^2 N(0)\int N(\xi_r)\delta(\xi_r)eVd\xi_r \\ &= \frac{4\pi e}{\hbar}|T|^2 N(0)^2 V \end{aligned}$$

or

$$G_{NN} = \frac{1}{V} = \frac{4\pi e}{\hbar} |T|^2 N(0)^2$$

## Superconductor-Insulator-Superconductor Junction

In order to describe superconductor-insulator superconductor junction, we express the tunneling Hamiltonian in terms of Bogoliubov-Valatin operators.

$$c_l^+ = u_l^* \gamma_l^+ + v_l \gamma_l$$
$$c_l = u_l \gamma_l + u_l^* \gamma_l^+$$

Similar relations hold for $c_r^+$ and $c_r$. Then the tunneling Hamiltonian becomes

$$H_t = \sum_{lr} T_{rl} \left[ \left( u_r^* \gamma_r^+ + u_r \gamma_r \right) \left( u_l \gamma_l + v_l^* \gamma_l^+ \right) + hc \right]$$
$$= \sum_{lr} T_{rl} \left[ \left( u_r^* u_l \gamma_r^+ \gamma_l + u_r^* u_l^* \gamma_r^+ \gamma_l^+ + u_r u_l \gamma_l \gamma_r + u_r u_l^* \gamma_r \gamma_l^+ \right) + hc \right]$$

We note that $|\phi_l>$ and $|\phi_R>$ represent the superconducting ground states on the left and right of the insulating material. We denote the excited states with quantum numbers $l$ and $r$ as

$$|l> = \gamma_e^+ |\phi_l>$$
$$|r> = \gamma_r^+ |\phi_r>$$

A combined state with one excitation $l$ in $L$ and no excitation in $R$ is written as

$$\gamma_l^+ |\phi_l> |\phi_r> = |l, \phi_r>$$

Similarly,

$$\gamma_r^+ |\phi_r> |\phi_l> = |\mathrm{r}, \phi_l>$$

for an excited stated in $R$ and no excitation in $L$. The matrix elements that lead to charge transfer from $L$ to $R$ consist of the following four processes

$$T_{rl} u_l^* u_r < \phi_l, r | \gamma_r^+ \gamma_l | l, \phi_r > + T_{lr}^* v_r^* v_l < \phi_l, r | \gamma_l \gamma_r^+ | l, \phi_r >$$

$$T_{rl} u_r^* v_l^* < l, r | \gamma_r^+ \gamma_l^+ | \phi_l, \phi_R > + T_{lr}^* v_r^* u_l^* < l, r | \gamma_l^+ \gamma_r^+ | \phi_l, \phi_r >$$

$$T_{lr}^* u_r v_l < \phi_e, \phi_r | \gamma_l \gamma_r | l, r > + T_{rl} v_r u_l < \phi_l, \phi_r | \gamma_r \gamma_l | l, r >$$

$$T_{lr}^* u_r u_l^* < l, \phi_r | \gamma_l^+ \gamma_r | \phi_l, r > + T_{rl} u_r v_l^* < l, \phi_r | \gamma_r \gamma_l^+ | \phi_l, r >$$

The corresponding energy differences of the initial and final states are given by

$$\Delta E = \varepsilon_l - (eV + \varepsilon_r)$$
$$\Delta E = eV - (\varepsilon_l + \varepsilon_r)$$
$$\Delta E = -eV + (\varepsilon_l + \varepsilon_r)$$
$$\Delta E = \varepsilon_r - (\varepsilon_l - eV)$$

Using

$$< \gamma_n^+ \gamma_{n'} >= f_n \delta_{nn'}$$
$$< \gamma_n \gamma_{n'}^+ >= (1 - f_n) \delta_{nn'}$$

And

$$< \gamma_n \gamma_{n'} >=< \gamma_n^+ \gamma_{n'}^+ >= 0$$

for thermal averaging $< H_t^+ H_t >$, and noting that the $u's$ and $v's$ drop out, we have the transition rates for the four processes

$$P_{L \to R}^{qb} = 2\frac{2\pi}{\hbar}\sum_{lr} |T_{lr}|^2 f_l (1 - f_r) \delta(\varepsilon_l - \varepsilon_r - eV)$$
$$P_{R-L}^{qb} = 2\frac{2\pi}{\hbar}\sum_{lr} |T_{lr}|^2 (1 - f_l) f_r \delta(\varepsilon_r - \varepsilon_l + eV)$$
$$P^{pb} = 2\frac{2\pi}{\hbar}\sum_{lr} |T_{lr}|^2 (1 - f_l)(1 - f_r) \delta(-\varepsilon_r - \varepsilon_l + eV)$$

And

$$P^r = 2\frac{2\pi}{\hbar}\sum_{lr} |T_{lr}|^2 f_e f_r \delta(\varepsilon_l + \varepsilon_r - eV)$$

Now the total quasiparticle current, $I^{qp}$, is

$$I^{qp} = I^{qp} = e\left[ P_{L-R}^{qp} - P_{R-L}^{qb} \right]$$
$$= \frac{4\pi e}{\hbar}\sum_{lr} |T_{lr}|^2 \left[ f_l (1 - f_r) \delta(\varepsilon_l - \varepsilon_r - eV) - (1 - f_l) f_r \delta(\varepsilon_r - \varepsilon_l + eV) \right]$$
$$= \frac{4\pi e}{\hbar}\sum_{lr} |T_{lr}|^2 \left[ (f_l - f_r) \delta(\varepsilon_l - \varepsilon_r - eV) \right]$$

Similarly, the pair breaking and recombination terms result in

$$I^p = \frac{4\pi e}{\hbar}\sum_{lr} |T_{lr}|^2 \left[ (1 - f_r)(1 - f_r) \delta(-\varepsilon_r - \varepsilon_l + eV) - f_l f_r \delta(\varepsilon_l + \varepsilon_r - eV) \right]$$

$$= \frac{4\pi e}{\hbar}\sum_{lr}|T_{lr}|^2\left[(1-f_l-f_r+f_1f_r)\delta(-\varepsilon_r-\varepsilon_l+eV)-f_1f_r\delta(\varepsilon_l+\varepsilon_r-eV)\right]$$

$$= \frac{4\pi e}{\hbar}\sum_{lr}|T_{lr}|^2(1-f_l-f_r)\delta(\varepsilon_l+\varepsilon_r-eV)$$

In order to carry out the integration, we need to have the density of states of the superconducting states $N_s(\varepsilon)$. The superconducting density of states is most easily obtained by noting the equality

$$N_s(\varepsilon)d\varepsilon = N_n(\xi)d\xi$$

and taking $N_n(\xi) \simeq N_n(0)$, we get

$$\frac{N_s(\varepsilon)}{N_n(0)} = \frac{d\xi_k}{d\varepsilon_R} = \frac{d}{d\varepsilon_k}\left[\left[\varepsilon_k^2-\Delta^2\right]\right]$$

$$\frac{N_s(\varepsilon)}{N_n(0)} = \begin{cases} \dfrac{\varepsilon_k}{(\varepsilon_k^2-\Delta^2)} & \varepsilon > \Delta \\ 0 & \varepsilon < \Delta \end{cases}$$

Note the singularity at the gap edge in the superconducting density of states. Now the summation can be written as

$$\sum_{lr} \iint N_L(0)N_R(0)\frac{d\xi_l}{d\varepsilon_l}\frac{d\xi_r}{d\varepsilon_r}d\varepsilon_r d\varepsilon_l$$

and the quasiparticle tunneling and pair breaking terms become, assuming energy independent matrix elements,

$$I^{qp} = \frac{4\pi e}{\hbar}|T|^2 N(0)^2 \iint \frac{\varepsilon_l}{(\varepsilon_l^2-\Delta^2)}\frac{\varepsilon_r}{(\varepsilon_r^2-\Delta^2)}\left[(f_l-f_r)\delta(\varepsilon_l-\varepsilon_r-eV)\right]d\varepsilon_l d\varepsilon_r$$

Integrating over $\varepsilon_r$ the presence of the $\delta$-function leads to, replacing $\varepsilon_l$ by $\varepsilon$

$$I^{qp} = \frac{4\pi e}{\hbar}|T|^2 N(0)^2 \iint \frac{|\varepsilon|}{(\varepsilon^2-\Delta^2)}\frac{|\varepsilon-eV|}{\left[(\varepsilon-eV^2)-\Delta^2\right]}\left[f(\varepsilon-eV)-f(\varepsilon)\right]d\varepsilon$$

where we have replaced $\varepsilon_l$ by $\varepsilon$. Similarly, for pair breaking/recombination tunneling current we get

$$I^{qp} = \frac{4\pi e}{\hbar}|T|^2 N(0)^2 \iint \left[1-f(-\varepsilon+eV)-f(\varepsilon)\right]\frac{|eV-\varepsilon|}{\left[(eV-\varepsilon)^2-\Delta^2\right]}\frac{|\varepsilon|}{\left[(\varepsilon^2-\Delta^2)\right]}d\varepsilon$$

# Josephson Effect

Josephson junction array chip developed by the National Bureau of Standards as a standard volt.

The Josephson effect is the phenomenon of supercurrent—i.e. a current that flows indefinitely long without any voltage applied—across a device known as a Josephson junction (JJ), which consists of two superconductors coupled by a weak link. The weak link can consist of a thin insulating barrier (known as a superconductor–insulator–superconductor junction, or S-I-S), a short section of non-superconducting metal (S-N-S), or a physical constriction that weakens the superconductivity at the point of contact (S-s-S).

The Josephson effect is an example of a macroscopic quantum phenomenon. It is named after the British physicist Brian David Josephson, who predicted in 1962 the mathematical relationships for the current and voltage across the weak link. The DC Josephson effect had been seen in experiments prior to 1962, but had been attributed to "super-shorts" or breaches in the insulating barrier leading to the direct conduction of electrons between the superconductors. The first paper to claim the discovery of Josephson's effect, and to make the requisite experimental checks, was that of Philip Anderson and John Rowell. These authors were awarded patents on the effects that were never enforced, but never challenged.

Before Josephson's prediction, it was only known that normal (i.e. non-superconducting) electrons can flow through an insulating barrier, by means of quantum tunneling. Josephson was the first to predict the tunneling of superconducting Cooper pairs. For this work, Josephson received the Nobel Prize in Physics in 1973. Josephson junctions have important applications in quantum-mechanical circuits, such as SQUIDs, superconducting qubits, and RSFQ digital electronics. The NIST standard for one volt is achieved by an array of 20,208 Josephson junctions in series.

## Applications

The electrical symbol for a Josephson junction.

Types of Josephson junction include the pi Josephson junction, varphi Josephson junction, long Josephson junction, and Superconducting tunnel junction. A "Dayem bridge" is a thin-film variant of the Josephson junction in which the weak link consists of a superconducting wire with dimensions on the scale of a few micrometres or less. The Josephson junction count of a device is used as a benchmark for its complexity. The Josephson effect has found wide usage, for example in the following areas:

- SQUIDs, or superconducting quantum interference devices, are very sensitive magnetometers that operate via the Josephson effect. They are widely used in science and engineering.
- In precision metrology, the Josephson effect provides an exactly reproducible conversion between frequency and voltage. Since the frequency is already defined precisely and practically by the caesium standard, the Josephson effect is used, for most practical purposes, to give the standard representation of a volt, the Josephson voltage standard. However, BIPM has not changed the official SI unit definition.
- Single-electron transistors are often constructed of superconducting materials, allowing use to be made of the Josephson effect to achieve novel effects. The resulting device is called a "superconducting single-electron transistor."
- The Josephson effect is also used for the most precise measurements of elementary charge in terms of the Josephson constant and von Klitzing constant which is related to the quantum Hall effect.
- RSFQ digital electronics is based on shunted Josephson junctions. In this case, the junction switching event is associated to the emission of one magnetic flux quantum $\frac{1}{2e}h$ that carries the digital information: the absence of switching is equivalent to *0*, while one switching event carries a *1*.
- Josephson junctions are integral in superconducting quantum computing as qubits such as in a flux qubit or others schemes where the phase and charge act as the conjugate variables.
- Superconducting tunnel junction detectors (STJs) may become a viable replacement for CCDs (charge-coupled devices) for use in astronomy and astrophysics in a few years. These devices are effective across a wide spectrum from ultraviolet to infrared, and also in x-rays. The technology has been tried out on the William Herschel Telescope in the SCAM instrument.
- Quiterons and similar superconducting switching devices.
- Josephson effect has also been observed in SHeQUIDs, the superfluid helium analog of a dc-SQUID.

## The Effect

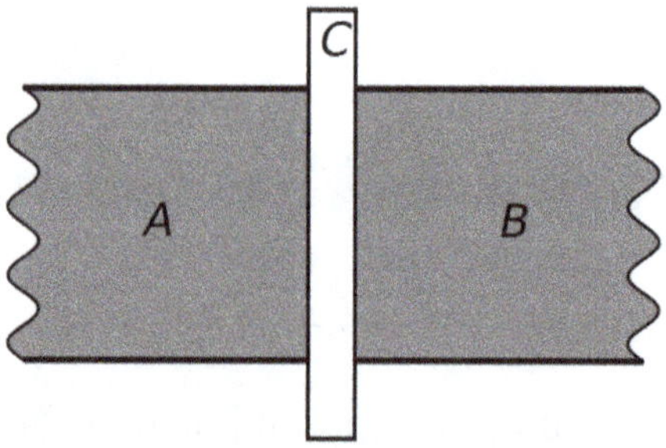


Diagram of a single Josephson junction. A and B represent superconductors, and C the weak link between them.

The basic equations governing the dynamics of the Josephson effect are

$$U(t) = \frac{\hbar}{2e}\frac{\partial \phi}{\partial t} \text{ (superconducting phase evolution equation)}$$

$$I(t) = I_c \sin(\phi(t)) \text{ (Josephson or weak-link current-phase relation)}$$

where $U(t)$ and $I(t)$ are the voltage across and the current through the Josephson junction, $\phi(t)$ is the "phase difference" across the junction (i.e., the difference in phase factor, or equivalently, argument, between the Ginzburg–Landau complex order parameter of the two superconductors composing the junction), and $I_c$ is a constant, the "critical current" of the junction. The critical current is an important phenomenological parameter of the device that can be affected by temperature as well as by an applied magnetic field. The physical constant $\frac{h}{2e}$ is the magnetic flux quantum, the inverse of which is the Josephson constant.

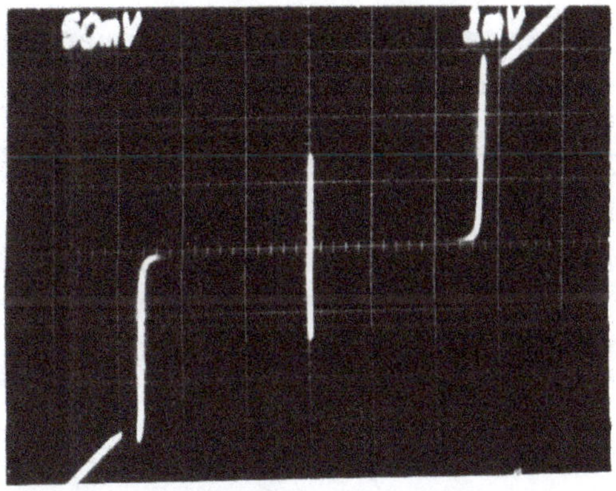


Typical I-V characteristic of a superconducting tunnel junction, a common kind of Josephson junction. The scale of the vertical axis is 50 μA and that of the horizontal one is 1 mV. The bar at $U = 0$ represents the DC Josephson effect, while the current at large values of $|U|$ is due to the finite value of the superconductor bandgap and not reproduced by the above equations.

The three main effects predicted by Josephson follow from these relations:

## The DC Josephson Effect

The DC Josephson effect is a direct current crossing the insulator in the absence of any external electromagnetic field, owing to tunneling. This DC Josephson current is proportional to the sine of the phase difference across the insulator, and may take values between $-I_c$ and $I_c$.

## The AC Josephson Effect

With a fixed voltage $U_{DC}$ across the junctions, the phase will vary linearly with time and the current will be an AC current with amplitude $I_c$ and frequency $\frac{2e}{h}U_{DC}$. The complete expression for the current drive $I_{ext}$ becomes:

$$I_{ext} = C_J \frac{dv}{dt} + I_J \sin\phi + \frac{V}{R}.$$

This means a Josephson junction can act as a perfect voltage-to-frequency converter.

## The Inverse AC Josephson Effect

If the phase takes the form $\phi(t)=\phi_0+n\omega t+a\sin(\omega t)$, the voltage and current will be

$U(t)=\frac{\hbar}{2e}\omega(n+a\cos(\omega t))$, and $I(t)=I_c\sum_{m=-\infty}^{\infty}J_m(a)\sin(\phi_0+(n+m)\omega t)$.

The DC components will then be

$U_{DC}=n\frac{\hbar}{2e}\omega$, and $I(t)=I_cJ_{-n}(a)\sin\phi_0$.

Hence, for distinct AC voltages, the junction may carry a DC current and the junction acts like a perfect frequency-to-voltage converter.

## Josephson Phase

The Josephson phase is the difference of the phases of the quantum mechanical wave function in two superconducting electrodes forming a Josephson junction.

If the macroscopic wave functions $\Psi_1$ and $\Psi_2$ in superconductors 1 and 2 are given by

$$\Psi_j=\sqrt{n_s}e^{i\theta_j},\forall j\in\{1,2\},$$

then the Josephson phase is defined by

$$\phi\overset{\text{def}}{=}\theta_2-\theta_1.$$

## Josephson Energy

The Josephson energy is the potential energy accumulated in a Josephson junction when a supercurrent flows through it. One can think of a Josephson junction as a non-linear inductance which accumulates (magnetic field) energy when a current passes through it. In contrast to real inductance, no magnetic field is created by a supercurrent in a Josephson junction — the accumulated energy is the Josephson energy.

For the simplest case the current-phase relation (CPR) is given by the first Josephson relation:

$$I_s=I_c\sin(\phi),$$

where $I_s$, is the supercurrent flowing through the junction, $I_c$, is the critical current, and $\phi$, is the Josephson phase. Imagine that initially at time $t=0$ the junction was in the ground state $\phi=0$ and finally at time $t$ the junction has the phase $\phi$. The work done on the junction (so the junction energy is increased by)

$$U=\int_0^t I_sV\,\mathrm{d}t=\frac{\Phi_0}{2\pi}\int_0^t I_s\frac{\mathrm{d}\phi}{\mathrm{d}t}\mathrm{d}t=\frac{\Phi_0}{2\pi}\int_0^\phi I_c\sin\phi\,\mathrm{d}\phi=\frac{\Phi_0I_c}{2\pi}(1-\cos\phi).$$

Here $E_J=\frac{\Phi_0I_c}{2\pi}$ sets the characteristic scale of the Josephson energy, and $1-\cos\phi$ sets its dependence on the phase $\phi$. The energy $U(\phi)$ accumulated inside the junction depends only on the current state of the junction, but not on history or velocities, i.e. it is a potential energy. Note, that $U(\phi)$ has a minimum equal to zero for the ground state $\phi = 2\pi n$, $n$ is any integer.

## Josephson Inductance

Imagine that the Josephson phase across the junction is $\phi_0$, and the supercurrent flowing through the junction is

$$I_0 = I_c \sin\phi_0.$$

(This is the same equation as above, except now we will look at small variations in $I_s$ and $\phi$ around the values $I_0$ and $\phi_0$.)

Imagine that we add little extra current (direct or alternative) $\delta_{I(t)} \ll I_c$ through the junction, and want to see how it reacts. The phase across the junction changes to become $\phi = \phi_0 + \delta_\phi$. One can write:

$$I_0 + \delta_I = I_c \sin(\phi_0 + \delta_\phi)$$

Assuming that $\delta_\phi$ is small, we make a Taylor expansion in the right hand side to arrive at

$$\delta_I = I_c \cos(\phi_0)\delta_\phi$$

The voltage across the junction (we use the 2nd Josephson relation) is

$$V = \frac{\Phi_0}{2\pi}\dot{\phi} = \frac{\Phi_0}{2\pi}(\underbrace{\dot{\phi_0}}_{=0} + \dot{\delta_\phi}) = \frac{\Phi_0}{2\pi}\frac{\dot{\delta_I}}{I_c\cos(\phi_0)}.$$

If we compare this expression with the expression for voltage across the conventional inductance

$$V = L\frac{\partial I}{\partial t},$$

we can define the so-called Josephson inductance

$$L_J(\phi_0) = \frac{\Phi_0}{2\pi I_c \cos(\phi_0)} = \frac{L_J(0)}{\cos(\phi_0)}.$$

One can see that this inductance is not constant, but depends on the phase $\phi_0$ across the junction. The typical value is given by $L_J(0)$ and is determined only by the critical current $I_c$. Note that, according to definition, the Josephson inductance can even become infinite or negative, if $\cos\phi_0 \le 0$.

One can also calculate the change in Josephson energy

$$\delta_{U(\phi_0)} = U(\phi) - U(\phi_0) = E_J(\cos(\phi_0) - \cos(\phi_0 + \delta_\phi)).$$

Making Taylor expansion for small $\delta_\phi$, we get

$$\approx E_J \sin(\phi_0)\delta_\phi = \frac{E_J \sin(\phi_0)}{I_c \cos\phi_0}\delta_I.$$

If we now compare this with the expression for increase of the inductance energy $\delta_{E_L} = LI\delta_I$, we again get the same expression for $L$.

Note, that although Josephson junction behaves like an inductance, there is no associated magnetic field. The corresponding energy is hidden inside the junction. The Josephson Inductance is also known as a Kinetic Inductance - the behaviour is derived from the kinetic energy of the charge carriers, not energy in a magnetic field.

As an alternative to the above approach to finding the Josephson Inductance, the equation for voltage across an inductor can be used (given by $V = L\frac{\partial I}{\partial t}$). By finding the derivative of the current with respect to time, and rearranging in the form of the inductance equation, inductance can be found.

Firstly, using the chain rule

$$\frac{\partial I}{\partial t} = \frac{\partial I}{\partial \phi}\frac{\partial \phi}{\partial t},$$

And from the Josephson junction equations

$$\frac{\partial I}{\partial \phi} = I_C cos(\phi),$$

$$\frac{\partial I}{\partial t} = \frac{2eV}{\hbar} = \frac{2\pi V}{\Phi_0},$$

Combining these three equations gives

$$\frac{\partial I}{\partial t} = \frac{I_c cos(\phi)2eV}{\hbar} = \frac{I_c cos(\phi)2\pi V}{\Phi_0},$$

and by rearranging to find in the form of $V = L\frac{\partial I}{\partial t}$

$$V = \frac{\hbar}{2eI_c cos(\phi)}\frac{\partial I}{\partial t}.$$

## Josephson Penetration Depth

The Josephson penetration depth characterizes the typical length on which an externally applied magnetic field penetrates into the long Josephson junction. Josephson penetration depth is usually denoted as $\lambda_J$ and is given by the following expression (in SI):

$$\lambda_J = \sqrt{\frac{\Phi_0}{2\pi\mu_0 d' j_c}},$$

where $\Phi_0$ is the magnetic flux quantum,       is the critical current density (A/m²), and $d'$ characterizes the inductance of the superconducting electrodes

$$d' = d_I + \lambda_1 \coth\left(\frac{d_1}{\lambda_1}\right) + \lambda_2 \coth\left(\frac{d_2}{\lambda_2}\right),$$

where $d_I$ is the thickness of the Josephson barrier (usually insulator), $d_I$ and $d_2$ are the thicknesses of superconducting electrodes, and $\lambda_1$ and $\lambda_2$ are their London penetration depths.

## Voltage-to-curvature Converter

A geometric potential from the kinetic term of a quantum superconducting condensate can be derived. It has the form $-\frac{\hbar^2}{24m^*}R^{(3d)}$, where $\hbar$ is Plack's constant, $m^*$ is the effective mass of the superconducting bosons and $R^{(3d)}$ is the three dimensional scalar curvature. The three dimensional scalar curvature is related to the complete four-dimensional scalar curvature $R$ of space-time by the relation $R^{(3d)} = \frac{3}{4}R$. At a Josephson junction an energy conservation relation applies and points at the possibility to transform electric energy into geometric (gravitational) field energy, that is curvature of space-time $\frac{\hbar^2}{24m^*}R^{(3d)} = 2eU$, where $e$ is the electron charge and $U$ is the voltage drop between the Josephson junction contacts. In effect, the Josephson junction can act as a voltage-to-curvature converter.

## Pair Tunneling, Modified Bogoliubov-Valatin Transformation and the Josephson Effects

In the following we will concentrate on $R(t-\tau)$ which leads to Cooper pair tunneling. In the following we will modify Bogoliubov and Valatin transformation to allow for pair number to change while conserving charge,

$$\begin{aligned}
\gamma^+_{e\vec{k}\uparrow} &= u_{\vec{k}} C^+_{\vec{k}\uparrow} - v_{\vec{k}} S^* C_{-\vec{k}\downarrow} \\
\gamma^+_{h\vec{k}\uparrow} &= u_{\vec{k}} S C^+_{\vec{k}\uparrow} - v_{\vec{k}} C_{-\vec{k}\downarrow} = S\gamma^+_{e\vec{k}\uparrow} \\
\gamma^+_{e\vec{k}\downarrow} &= u_{\vec{k}} C^+_{\vec{k}\downarrow} + v_{\vec{k}} S^* C_{\vec{k}\uparrow} \\
\gamma^+_{n\vec{k}\downarrow} &= u_{\vec{k}} S C^+_{-\vec{k}\downarrow} + v_{\vec{k}} C_{\vec{k}\uparrow} = S\gamma^+_{e\vec{k}\downarrow}
\end{aligned}$$

where $S^*$ adds a Cooper pair and $S$ destroys one Cooper pair from the condensate. The subscripts $e$ and $h$ refer to electron and hole respectively.

The inverse operators are

$$C^{+}_{\vec{k}\uparrow} = u_{\vec{k}}\gamma^{+}_{e\vec{k}\uparrow} + v_{\vec{k}}\gamma_{h\vec{k}\downarrow}$$
$$C^{+}_{\vec{k}\uparrow} = u_{k}\gamma^{+}_{e\vec{k}\uparrow} - v_{\vec{k}}\gamma_{h\vec{k}\downarrow}$$

Now the total energy of the superconductor can be written as

$$E = \sum_{\vec{k}} \varepsilon_{\vec{k}}\gamma^{+}_{\vec{k}}\gamma_{\vec{k}} + \mu N$$

where sum runs for all excitations with $k > 0$. The energy needed to make an electron-like excitation is

$$\varepsilon_{e\vec{k}} = \varepsilon_{\vec{k}} + \mu$$

while the energy needed for a hole-like excitation is

$$\varepsilon_{h\vec{k}} = -\mu + \varepsilon_{\vec{k}}$$

In the interaction representation the quasiparticle operators are given by

$$\gamma_{el}(t) = \gamma_{el}(0)e^{-i(\varepsilon_l+\mu_l)t/\hbar}$$

and

$$\gamma_{hl}(t) = \gamma_{hl}(0)e^{-i(\varepsilon_l-\mu_l)t/\hbar}$$

For calculating the pair tunneling current from $R(t-\tau)$ we note that we have a term like

$$u_l v_l u_r v_r \left[<\gamma_{hl|}(t)\gamma^{+}_{el|}(\tau)>_0 <\gamma^{+}_{hr|}(t)\gamma_{er|}(\tau)>_0 - <\gamma^{+}_{el|}(\tau)\gamma_{hl|}(t)>_0 <\gamma^{+}_{er|}(\tau)\gamma_{hr|}(t)>_0\right]$$

If the bias across the junction is such that

$$\mu_l - \mu_r = eV$$

then we can write the above expression as

$$u_l v_l u_r v_r S^{*}_l S_r e^{-i(\varepsilon_l-\varepsilon_r+eV)(t-\tau)/\hbar}\left[<\gamma_{el|}(0)\gamma^{+}_{el|}(0)>_0 \; <\gamma^{+}_{er|}(0)\gamma_{er|}(0)>_0 - <\gamma^{+}_{el|}(0)\gamma_{el|}(0)>_0 \; <\gamma_{er|}(0)\gamma^{+}_{er|}(0)>_0\right]$$

If the phase difference between the Cooper pairs on the left and on right of the junction is $\Phi$ then

$$S^{*}_l S_r = e^{-i\Phi}$$

Collecting all the terms conserving change, energy and spin and carrying out $\tau$ -integration, we get the Josephson current as

$$I_J = -\frac{e}{\hbar} I_m \left( \sum_{l_r r_r} \frac{4|T_{lr}|^2 u_l v_l u_r v_r e^{-i\Phi}(f_r - f_l)}{\varepsilon_l - \mu_l - eV + i\hbar\eta} + cc \right)$$

where the summation is only for electron-like branches on the left $l_>$ and on the right $r_>$. Changing the sum into integration and noting that the integral has a principal part and a residue part, we can write,

$$I_J = I_1 \sin\Phi + I_2 \cos\Phi$$

Where

$$I_1 = \frac{2e|T|^2}{\hbar} N_L(0) N_R(0) P \int_{-\infty}^{\infty} d\varepsilon_1 \int_{-\infty}^{\infty} d\varepsilon_2 \frac{\Delta_1 \Delta_2}{\varepsilon_1 \varepsilon_2} \frac{f(\varepsilon_1) - f(\varepsilon_2)}{\varepsilon_1 - \varepsilon_2 + eV}$$

and

$$I_2 = \frac{2\pi e|T|^2}{\hbar} N_L(0) N_R(0) \int_{-\infty}^{\infty} d\varepsilon \frac{\Delta_1 \Delta_2}{(\varepsilon - eV)\varepsilon} [f(\varepsilon - eV) - f(\varepsilon)]$$

The pair tunneling current $I_1 \sin\Phi$ is known as the Josephson supercurrrent and the effect is known as the Josephson effect. The Josephson supercurrent is finite even when $V = 0$.

## References


- B. D. Josephson, "Possible new effects in superconductive tunnelling," Physics Letters 1, 251 (1962), doi:10.1016/0031-9163(62)91369-0
- Barone, A.; Paterno, G. (1982). Physics and Applications of the Josephson Effect. New York: John Wiley & Sons. ISBN 0-471-01469-9
- C. A. Hamilton, R. L. Kautz, R. L. Steiner, and F. L. Lloyd, "A practical Josephson voltage standard at 1 V," IEEE Electron Device Letters 6, 623 (1985), doi:10.1109/EDL.1985.26253
- P. W. Anderson; A. H. Dayem (1964). "Radio-frequency effects in superconducting thin film bridges". Phys. Rev. Lett. 13 (6): 195. doi:10.1103/PhysRevLett.13.195
- J. R. Tucker, "Quantum limited detection in tunnel junction mixers," IEEE Journal of Quantum Electronics 15, 1234 (1979), doi:10.1109/JQE.1979.1069931
- Ficken, F. A. (1939-01-01). "The Riemannian and Affine Differential Geometry of Product-Spaces". Annals of Mathematics. 40 (4): 892–913. doi:10.2307/1968900

# 6

# Electrical and Magnetic Conductivity in Metals

A good value of resistivity of a metal at room temperature is 10-6 ohm cm. At high temperatures, heat capacity tends to a constant value while at low temperatures the heat capacity is a combination of a linear term and a cubic term. The topics discussed in the chapter are of great importance to broaden the existing knowledge on superconductivity.

## Metal

A metal is a material (an element, compound, or alloy) that is typically hard, opaque, shiny, and has good electrical and thermal conductivity. Metals are generally malleable—that is, they can be hammered or pressed permanently out of shape without breaking or cracking—as well as fusible (able to be fused or melted) and ductile (able to be drawn out into a thin wire). About 91 of the 118 elements in the periodic table are metals, the others are nonmetals or metalloids. Some elements appear in both metallic and non-metallic forms.

Astrophysicists use the term "metal" to collectively describe all elements other than hydrogen and helium, the simplest two, in a star. The star fuses smaller atoms, mostly hydrogen and helium, to make larger ones over its lifetime. In that sense, the metallicity of an object is the proportion of its matter made up of all heavier chemical elements, not just traditional metals.

Many elements and compounds that are not normally classified as metals become metallic under high pressures; these are formed as metallic allotropes of non-metals.

### Structure and Bonding

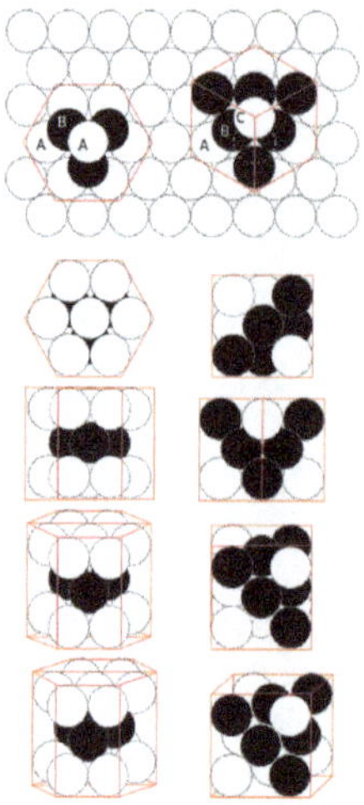

hcp and fcc close-packing of spheres.

The atoms of metallic substances are typically arranged in one of three common crystal structures, namely body-centered cubic (bcc), face-centered cubic (fcc), and hexagonal close-packed (hcp). In bcc, each atom is positioned at the center of a cube of eight others. In fcc and hcp, each atom is surrounded by twelve others, but the stacking of the layers differs. Some metals adopt different structures depending on the temperature.

Atoms of metals readily lose their outer shell electrons, resulting in a free flowing cloud of electrons within their otherwise solid arrangement. This provides the ability of metallic substances to easily transmit heat and electricity. While this flow of electrons occurs, the solid characteristic of the metal is produced by electrostatic interactions between each atom and the electron cloud. This type of bond is called a metallic bond.

## Properties

### Chemical

Metals are usually inclined to form cations through electron loss, reacting with oxygen in the air to form oxides over various timescales (iron rusts over years, while potassium burns in seconds). Examples:

$$4\,Na + O_2 \rightarrow 2\,Na_2O \text{ (sodium oxide)}$$

$$2\,Ca + O_2 \rightarrow 2\,CaO \text{ (calcium oxide)}$$

$$4\,Al + 3\,O_2 \rightarrow 2\,Al_2O_3 \text{ (aluminium oxide).}$$

The transition metals (such as iron, copper, zinc, and nickel) are slower to oxidize because they form a passivating layer of oxide that protects the interior. Others, like palladium, platinum and gold, do not react with the atmosphere at all. Some metals form a barrier layer of oxide on their surface which cannot be penetrated by further oxygen molecules and thus retain their shiny appearance and good conductivity for many decades (like aluminium, magnesium, some steels, and titanium). The oxides of metals are generally basic, as opposed to those of nonmetals, which are acidic. Exceptions are largely oxides with very high oxidation states such as $CrO_3$, $Mn_2O_7$, and $OsO_4$, which have strictly acidic reactions.

Painting, anodizing or plating metals are good ways to prevent their corrosion. However, a more reactive metal in the electrochemical series must be chosen for coating, especially when chipping of the coating is expected. Water and the two metals form an electrochemical cell, and if the coating is less reactive than the coatee, the coating actually *promotes* corrosion.

### Physical

Metals in general have high electrical conductivity, high thermal conductivity, and high density. Typically they are malleable and ductile, deforming under stress without cleaving. In terms of optical properties, metals are shiny and lustrous. Sheets of metal beyond a few micrometres in thickness appear opaque, but gold leaf transmits green light.

Although most metals have higher densities than most nonmetals, there is wide variation in their densities, lithium being the least dense solid element and osmium the densest. The alkali and alkaline earth metals in groups I A and II A are referred to as the light metals because they have low density, low hardness, and low melting points. The high density of most metals is due to the tightly packed crystal lattice of the metallic structure. The strength of metallic bonds for different metals reaches a maximum around the center of the transition metal series, as those elements have large amounts of delocalized electrons in tight binding type metallic bonds. However, other factors (such as atomic radius, nuclear charge, number of bonds orbitals, overlap of orbital energies and crystal form) are involved as well.

Gallium crystals.

## Electrical

The electrical and thermal conductivities of metals originate from the fact that their outer electrons are delocalized. This situation can be visualized by seeing the atomic structure of a metal as a collection of atoms embedded in a sea of highly mobile electrons. The electrical conductivity, as well as the electrons' contribution to the heat capacity and heat conductivity of metals can be calculated from the free electron model, which does not take into account the detailed structure of the ion lattice.

When considering the electronic band structure and binding energy of a metal, it is necessary to take into account the positive potential caused by the specific arrangement of the ion cores—which is periodic in crystals. The most important consequence of the periodic potential is the formation of a small band gap at the boundary of the Brillouin zone. Mathematically, the potential of the ion cores can be treated by various models, the simplest being the nearly free electron model.

## Mechanical

Mechanical properties of metals include ductility, i.e. their capacity for plastic deformation. Reversible elastic deformation in metals can be described by Hooke's Law for restoring forces, where the stress is linearly proportional to the strain. Forces larger than the elastic limit, or heat, may cause a permanent (irreversible) deformation of the object, known as plastic deformation or plasticity. This irreversible change in atomic arrangement may occur as a result of:

- The action of an applied force (or work). An applied force may be tensile (pulling) force, compressive (pushing) force, shear, bending or torsion (twisting) forces.

- A change in temperature (heat). A temperature change may affect the mobility of the structural defects such as grain boundaries, point vacancies, line and screw dislocations, stacking faults and twins in both crystalline and non-crystalline solids. The movement or displacement of such mobile defects is thermally activated, and thus limited by the rate of atomic diffusion.

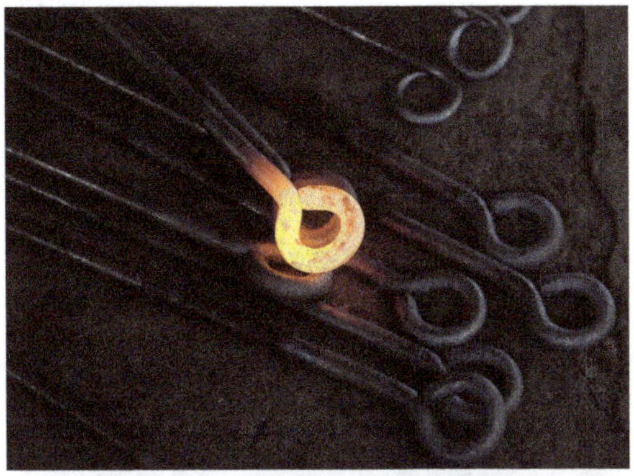

Hot metal work from a blacksmith.

Viscous flow near grain boundaries, for example, can give rise to internal slip, creep and fatigue in metals. It can also contribute to significant changes in the microstructure like grain growth and localized densification due to the elimination of intergranular porosity. Screw dislocations may slip in the direction of any lattice plane containing the dislocation, while the principal driving force for "dislocation climb" is the movement or diffusion of vacancies through a crystal lattice.

In addition, the nondirectional nature of metallic bonding is also thought to contribute significantly to the ductility of most metallic solids. When the planes of an ionic bond slide past one another, the resultant change in location shifts ions of the same charge into close proximity, resulting in the cleavage of the crystal; such shift is not observed in covalently bonded crystals where fracture and crystal fragmentation occurs.

## Alloys

An alloy is a mixture of two or more elements in which the main component is a metal. Most pure metals are either too soft, brittle or chemically reactive for practical use. Combining different ratios of metals as alloys modifies the properties of pure metals to produce desirable characteristics. The aim of making alloys is generally to make them less brittle, harder, resistant to corrosion, or have a more desirable color and luster. Of all the metallic alloys in use today, the alloys of iron (steel, stainless steel, cast iron, tool steel, alloy steel) make up the largest proportion both by quantity and commercial value. Iron alloyed with various proportions of carbon gives low, mid and high carbon steels, with increasing carbon levels reducing ductility and toughness. The addition of silicon will produce cast irons, while the addition of chromium, nickel and molybdenum to carbon steels (more than 10%) results in stainless steels.

Other significant metallic alloys are those of aluminium, titanium, copper and magnesium. Copper alloys have been known since prehistory—bronze gave the Bronze Age its name—and have many applications today, most importantly in electrical wiring. The alloys of the other three metals have been developed relatively recently; due to their chemical reactivity they require electrolytic extraction processes. The alloys of aluminium, titanium and magnesium are valued for their high strength-to-weight ratios; magnesium can also provide electromagnetic shielding. These materials are ideal for situations where high strength-to-weight ratio is more important than material cost, such as in aerospace and some automotive applications.

Alloys specially designed for highly demanding applications, such as jet engines, may contain more than ten elements.

## Categories

### Base Metal

Zinc, a base metal, reacting with an acid.

In chemistry, the term *base metal* is used informally to refer to a metal that is easily oxidized or corroded, and reacts easily with dilute hydrochloric acid (HCl) to form hydrogen. Examples include iron, nickel, lead and zinc. Copper is considered a base metal as it is oxidized relatively easily, although it does not react with HCl. Base metal is commonly used in opposition to noble metal.

In alchemy, a *base metal* was a common and inexpensive metal, as opposed to precious metals, mainly gold and silver. A longtime goal of the alchemists was the transmutation of base metals into precious metals.

In numismatics, coins in the past derived their value primarily from the precious metal content. Most modern currencies are fiat currency, allowing the coins to be made of base metal.

### Ferrous Metal

The term "ferrous" is derived from the Latin word meaning "containing iron". This can include pure iron, such as wrought iron, or an alloy such as steel. Ferrous metals are often magnetic, but not exclusively.

### Noble Metal

*Noble metals* are metals that are resistant to corrosion or oxidation, unlike most base metals. They tend to be precious metals, often due to perceived rarity. Examples include gold, platinum, silver, rhodium, iridium and palladium.

## Precious Metal

A gold nugget.

A *precious metal* is a rare metallic chemical element of high economic value.

Chemically, the precious metals are less reactive than most elements, have high luster and high electrical conductivity. Historically, precious metals were important as currency, but are now regarded mainly as investment and industrial commodities. Gold, silver, platinum and palladium each have an ISO 4217 currency code. The best-known precious metals are gold and silver. While both have industrial uses, they are better known for their uses in art, jewelry, and coinage. Other precious metals include the platinum group metals: ruthenium, rhodium, palladium, osmium, iridium, and platinum, of which platinum is the most widely traded.

The demand for precious metals is driven not only by their practical use, but also by their role as investments and a store of value. Palladium was, as of summer 2006, valued at a little under half the price of gold, and platinum at around twice that of gold. Silver is substantially less expensive than these metals, but is often traditionally considered a precious metal for its role in coinage and jewelry.

## Heavy Metal

A heavy metal is any relatively dense metal or metalloid. More specific definitions have been proposed, but none have obtained widespread acceptance. Some heavy metals have niche uses, or are notably toxic; some are essential in trace amounts.

## Extraction

Metals are often extracted from the Earth by means of mining ores that are rich sources of the requisite elements, such as bauxite. Ore is located by prospecting techniques, followed by the exploration and examination of deposits. Mineral sources are generally divided into surface mines, which are mined by excavation using heavy equipment, and subsurface mines.

Once the ore is mined, the metals must be extracted, usually by chemical or electrolytic reduction. Pyrometallurgy uses high temperatures to convert ore into raw metals, while hydrometallurgy employs aqueous chemistry for the same purpose. The methods used depend on the metal and their contaminants.

When a metal ore is an ionic compound of that metal and a non-metal, the ore must usually be smelted—heated with a reducing agent—to extract the pure metal. Many common metals, such as iron, are smelted using carbon as a reducing agent. Some metals, such as aluminium and sodium, have no commercially practical reducing agent, and are extracted using electrolysis instead.

Sulfide ores are not reduced directly to the metal but are roasted in air to convert them to oxides.

## Recycling

Demand for metals is closely linked to economic growth. During the 20th century, the variety of metals uses in society grew rapidly. Today, the development of major nations, such as China and India, and advances in technologies, are fuelling ever more demand. The result is that mining activities are expanding, and more and more of the world's metal stocks are above ground in use, rather than below ground as unused reserves. An example is the in-use stock of copper. Between 1932 and 1999, copper in use in the US rose from 73g to 238g per person.

Metals are inherently recyclable, so in principle, can be used over and over again, minimizing these negative environmental impacts and saving energy. For example, 95% of the energy used to make aluminium from bauxite ore is saved by using recycled material. Levels of metals recycling are generally low. In 2010, the International Resource Panel, hosted by the United Nations Environment Programme (UNEP) published reports on metal stocks that exist within society and their recycling rates.

The report authors observed that the metal stocks in society can serve as huge mines above ground. They warned that the recycling rates of some rare metals used in applications such as mobile phones, battery packs for hybrid cars and fuel cells are so low that unless future end-of-life recycling rates are dramatically stepped up these critical metals will become unavailable for use in modern technology.

## Metallurgy

Metallurgy is a domain of materials science that studies the physical and chemical behavior of metallic elements, their intermetallic compounds, and their mixtures, which are called alloys.

## Applications

Some metals and metal alloys possess high structural strength per unit mass, making them useful materials for carrying large loads or resisting impact damage. Metal alloys can be engineered to have high resistance to shear, torque and deformation. However the same metal can also be vulnerable to fatigue damage through repeated use or from sudden stress failure when a load capacity is exceeded. The strength and resilience of metals has led to their frequent use in high-rise building and bridge construction, as well as most vehicles, many appliances, tools, pipes, non-illuminated signs and railroad tracks.

The two most commonly used structural metals, iron and aluminium, are also the most abundant metals in the Earth's crust.

Metals are good conductors, making them valuable in electrical appliances and for carrying an electric current over a distance with little energy lost. Electrical power grids rely on metal cables to distribute electricity. Home electrical systems, for the most part, are wired with copper wire for its good conducting properties.

The thermal conductivity of metal is useful for containers to heat materials over a flame. Metal is also used for heat sinks to protect sensitive equipment from overheating.

The high reflectivity of some metals is important in the construction of mirrors, including precision astronomical instruments. This last property can also make metallic jewelry aesthetically appealing.

Some metals have specialized uses; radioactive metals such as uranium and plutonium are used in nuclear power plants to produce energy via nuclear fission. Mercury is a liquid at room temperature and is used in switches to complete a circuit when it flows over the switch contacts. Shape memory alloy is used for applications such as pipes, fasteners and vascular stents.

Metals can be doped with foreign molecules—organic, inorganic, biological and polymers. This doping entails the metal with new properties that are induced by the guest molecules. Applications in catalysis, medicine, electrochemical cells, corrosion and more have been developed.

## Trade

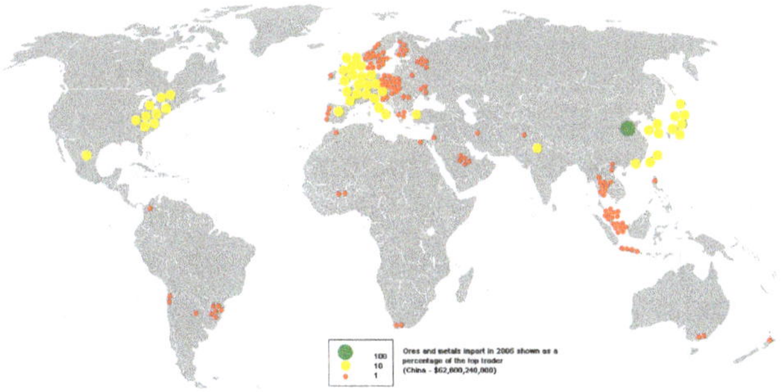


World metal and ore imports in 2005.

The World Bank reports that China was the top importer of ores and metals in 2005 followed by the United States and Japan.

## History

The nature of metals has fascinated humans for many centuries, because these materials provided people with tools of unsurpassed properties both in war and in their preparation and processing. Pure gold and silver were known to humans since the Stone Age. Lead and silver were fused from their ores as early as the fourth millennium BC.

Ancient Latin and Greek writers such as Theophrastus, Pliny the Elder in his *Natural History*, or Pedanius Dioscorides, did not try to classify metals. The ancient Europeans never attained the concept "metal" as a distinct elementary substance with fixed, characteristic chemical and physical properties. Following Empedocles, all substances within the sublunary sphere were assumed to vary in their constituent classical elements of earth, water, air and fire. Following the Pythagoreans, Plato assumed that these elements could be further reduced to plane geometrical shapes (triangles and squares) bounding space and relating to the regular polyhedra in the sequence earth:cube, water:icosahedron, air:octahedron, fire:tetrahedron. However, this philosophical extension did not become as popular as the simple four elements, after it was rejected by Aristotle. Aristotle also rejected the atomic theory of Democritus, since he classified the implied existence of a vacuum necessary for motion as a contradiction (a vacuum implies nonexistence, therefore cannot exist). Aristotle did, however, introduce underlying antagonistic qualities (or forces) of dry vs. wet and cold vs. heat into the composition of each of the four elements. The word "metal" originally meant "mines" and only later gained the general meaning of products from materials obtained in mines. In the first centuries A.D. a relation between the planets and the existing metals was assumed as Gold:Sun, Silver:Moon, Electrum:Jupiter, Iron:Mars, Copper:Venus, Tin:Mercury, Lead:Saturn. After electrum was determined to be a combination of silver and gold, the relations Tin:Jupiter and Mercury:Mercury were substituted into the previous sequence.

Arabic and medieval alchemists believed that all metals, and in fact, all sublunar matter, were composed of the principle of sulfur, carrying the combustible property, and the principle of mercury, the mother of all metals and carrier of the liquidity or fusibility, and the volatility properties. These principles were not necessarily the common substances sulfur and mercury found in most laboratories. This theory reinforced the belief that the all metals were destined to become gold in the bowels of the earth through the proper combinations of heat, digestion, time, and elimination of contaminants, all of which could be developed and hastened through the knowledge and methods of alchemy. Paracelsus added the third principle of salt, carrying the nonvolatile and incombustible properties, in his *tria prima* doctrine. These theories retained the four classical elements as underlying the composition of sulfur, mercury and salt.

The first systematic text on the arts of mining and metallurgy was *De la Pirotechnia* by Vannoccio Biringuccio, which treats the examination, fusion, and working of metals. Sixteen years later, Georgius Agricola published *De Re Metallica* in 1555, a clear and complete account of the profession of mining, metallurgy, and the accessory arts and sciences, as well as qualifying as the greatest treatise on the chemical industry through the sixteenth century. He gave the following description of a metal in his *De Natura Fossilium* (1546).

Metal is a mineral body, by nature either liquid or somewhat hard. The latter may be melted by the heat of the fire, but when it has cooled down again and lost all heat, it becomes hard again and resumes its proper form. In this respect it differs from the stone which melts in the fire, for although the latter regain its hardness, yet it loses its pristine form and properties. Traditionally there are six different kinds of metals, namely gold, silver, copper, iron, tin and lead. There are really others, for quicksilver is a metal, although the Alchemists disagree with us on this subject, and bismuth is also. The ancient Greek writers seem to have been ignorant of bismuth, wherefore Ammonius rightly states that there are many species of metals, animals, and plants which are unknown to us. Stibium when smelted in the crucible and refined has as much right to be regarded as a proper metal as is accorded to lead by writers. If when smelted, a certain portion be added to tin, a bookseller's alloy is produced from which the type is made that is used by those who print books on paper. Each metal has its own form which it preserves when separated from those metals which were mixed with it. Therefore neither electrum nor Stannum [not meaning our tin] is of itself a real metal, but rather an alloy of two metals. Electrum is an alloy of gold and silver, Stannum of lead and silver. And yet if silver be parted from the electrum, then gold remains and not electrum; if silver be taken away from Stannum, then lead remains and not Stannum. Whether brass, however, is found as a native metal or not, cannot be ascertained with any surety. We only know of the artificial brass, which consists of copper tinted with the colour of the mineral calamine. And yet if any should be dug up, it would be a proper metal. Black and white copper seem to be different from the red kind. Metal, therefore, is by nature either solid, as I have stated, or fluid, as in the unique case of quicksilver. But enough now concerning the simple kinds.

## Basic Properties Of Metals

Before delving into the properties of superconductors, it is important to give a brief account of the salient features of normal metals. In a non-interacting electron picture, the contribution of the elecrons to various properties can be calculated. This is worked out in standard texts on solid state physics and the summary of the results is given below.

## Electrical Conductivity

We recall that in the non-interacting-electrons description of metals, one considers electrons moving in a periodic potential of the lattice. In this "Bloch" picture, a perfect crystal is expected to have an infinite electrical conductivity. In real materials, a finite conductivity appears due to the inherent imperfections and defects. Additionally, at non-zero temperatures, lattice vibrations lead to a deviation from periodicity and contribute to electron scattering. In summary, one gets resistivity $\rho \sim$ T for T$\gg \Theta_D$ where $\Theta_D$ is the Debye temperature. On the other hand, for T $\ll \Theta_D$ one obtains $\rho \sim T^5$. The residual resistivity as $T \rightarrow 0$ decreases with decreasing amounts of impurities. A typical value of resistivity of a good metal at room temperature is of the order of $10^{-6}$ ohm cm.

As an example, the variation of the resistivity with temperature for silver is shown in the figure below

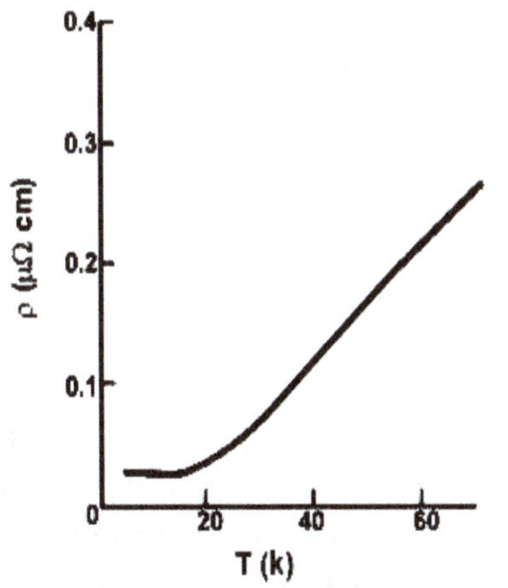


**Schematic variation of resistivity of a metal with temperature**

The resistivity variation with temperature is shown for a normal metal. At high temperature, a linear variation of resistivity with temperature is seen while at low temperatures a residual resistivity is present which is linked to the presence of defects and impurities.

For ideal metals, the thermal conductivity due to electrons is given by $\frac{1}{3}v^2\tau c_v$ where $v^2$ is the mean square electronic speed, $\tau$ is the relaxation time, and $c_v$ is the specific heat capacity of the electrons. The ratio of the thermal conductivity $K$ to the electrical conductivity times temperature ($\sigma$T) of an electron gas is a universal constant (Lorentz number) and this is called Wiedmann-Franz law. The value of the Lorentz number is of the order of $10^{-8}\,\frac{\text{Wattohm}}{\text{K}^2}$ in typical metals.

### Heat Capacity

The temperature dependence of the specific heat capacity of an electron gas is given by $\frac{c_v}{k_B} = \gamma T = \frac{\pi^3}{3} g(\epsilon_F) k_B T$ where $k_B$ is the Boltzmann constant and $g(\epsilon_F)$ is the density-of-states at the Fermi level. The value of $\gamma$ in typical metals is of the order of $1\,\frac{\text{mJ}}{\text{moleK}^2}$. In the Debye model, the phonon contribution to the heat capacity per atom is given by $\frac{c_v}{k_B} = 9\left(\frac{T}{\Theta_D}\right)^3 \int_0^{\Theta_D/T} \frac{x^4 e^x dx}{(e^x - 1)^2}$

where $\Theta_D$ is the Debye temperature.

At high temperatures, it tends to a constant value while at low temperatures the heat capacity is a combination of a linear term (due to electrons) and a cubic term (due to lattice vibrations or phonons). When C/T is plotted as function of $T^2$, a straight line is obtained. The intercept on the y-axis gives information about the electron density of states while the slope provides information about characteristic energy associated with lattice vibrations.

In the low-temperature limit $T \ll \Theta_D$, the phonon contribution reduces to $\frac{c_v}{k_B} = \frac{12\pi^4}{5}\left(\frac{T}{\Theta_D}\right)^3$. In a typical metal, the low-temperature heat capacity has, therefore, a combination of a linear and a cubic term in temperature. The low-temperature heat capacity measurement serves as an important probe of the Fermi surface properties of the electron gas.

## Electrical Resistivity and Conductivity

Electrical resistivity (also known as resistivity, specific electrical resistance, or volume resistivity) is an intrinsic property that quantifies how strongly a given material opposes the flow of electric current. A low resistivity indicates a material that readily allows the flow of electric current. Resistivity is commonly represented by the Greek letter ρ (rho). The SI unit of electrical resistivity is the ohm-metre (Ω·m). As an example, if a 1 m × 1 m × 1 m solid cube of material has sheet contacts on two opposite faces, and the resistance between these contacts is 1 Ω, then the resistivity of the material is 1 Ω·m.

Electrical conductivity or specific conductance is the reciprocal of electrical resistivity, and measures a material's ability to conduct an electric current. It is commonly represented by the Greek letter σ (sigma), but κ (kappa) (especially in electrical engineering) or γ (gamma) are also occasionally used. Its SI unit is siemens per metre (S/m) and CGSE unit is reciprocal second ($s^{-1}$).

## Definition

### Resistors or Conductors with Uniform Cross-section

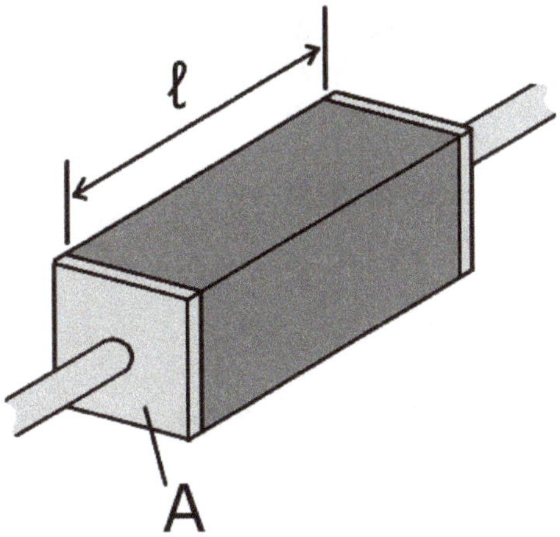


A piece of resistive material with electrical contacts on both ends.

Many resistors and conductors have a uniform cross section with a uniform flow of electric current, and are made of one material. In this case, the electrical resistivity $\rho$ is defined as:

$$\rho = R\frac{A}{\ell},$$

where

$R$ is the electrical resistance of a uniform specimen of the material

$\ell$ is the length of the piece of material

$A$ is the cross-sectional area of the specimen

The reason resistivity is defined this way is that it makes resistivity an *intrinsic property*, unlike resistance. All copper wires, irrespective of their shape and size, have approximately the same *resistivity*, but a long, thin copper wire has a much larger *resistance* than a thick, short copper wire. Every material has its own characteristic resistivity. For example, rubber has a far larger resistivity than copper.

In a hydraulic analogy, passing current through a high-resistivity material is like pushing water through a pipe full of sand—while passing current through a low-resistivity material is like pushing water through an empty pipe. If the pipes are the same size and shape, the pipe full of sand has higher resistance to flow. Resistance, however, is not *solely* determined by the presence or absence of sand. It also depends on the length and width of the pipe: short or wide pipes have lower resistance than narrow or long pipes.

The above equation can be transposed to get Pouillet's law (named after Claude Pouillet):

$$R = \rho\frac{\ell}{A}.$$

The resistance of a given material increases with length, but decreases with increasing cross-sectional area. From the above equations, resistivity has the SI unit "ohm metre" (Ω·m).

The formula $R = \rho\ell / A$ can be used to intuitively understand the meaning of a resistivity value. For example, if $A = 1\ \mathrm{m}^2$ $\ell = 1\ \mathrm{m}$ (forming a cube with perfectly conductive contacts on opposite faces), then the resistance of this element in ohms is numerically equal to the resistivity of the material it is made of in Ω·m.

Conductivity, σ, is defined as the inverse of resistivity:

$$\sigma = \frac{1}{\rho}.$$

Conductivity has SI units of "siemens per metre" (S/m).

## General Definition

The above definition was specific to resistors or conductors with a uniform cross-section, where current flows uniformly through them. A more basic and general definition starts from the fact that an electric field inside a material makes electric current flow. The electrical resistivity, $\rho$, is defined as the ratio of the electric field to the density of the current it creates:

$$\rho = \frac{E}{J},$$

where

$\rho$ is the resistivity of the conductor material,

$E$ is the magnitude of the electric field,

$J$ is the magnitude of the current density,

in which $E$ and $J$ are inside the conductor.

Conductivity is the inverse:

$$\sigma = \frac{1}{\rho} = \frac{J}{E}.$$

For example, rubber is a material with large $\rho$ and small $\sigma$—because even a very large electric field in rubber makes almost no current flow through it. On the other hand, copper is a material with small $\rho$ and large $\sigma$—because even a small electric field pulls a lot of current through it.

## Causes of Conductivity

## Band Theory Simplified

According to elementary quantum mechanics, electrons in an atom do not take on arbitrary energy values. Rather, electrons only occupy certain discrete energy levels in an atom or crystal; energies between these levels are impossible. When a large number of such allowed energy levels are spaced close together (in energy-space)—i.e. have similar (minutely differing) energies— we can talk about these energy levels together as an "energy band". There can be many such energy bands in a material, depending on the atomic number {number of electrons (if the atom is neutral)} and their distribution (besides external factors like environmental modification of the energy bands).

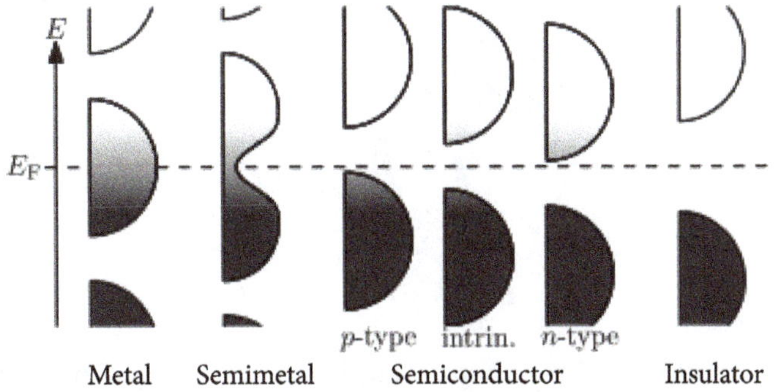


Filling of the electronic states in various types of materials at equilibrium. Here, height is energy while width is the density of available states for a certain energy in the material listed. The shade follows the Fermi–Dirac distribution (black = all states filled, white = no state filled). In metals and semimetals the Fermi level $E_F$ lies inside at least one band. In insulators and semiconductors the Fermi level is inside a band gap; however, in semiconductors the bands are near enough to the Fermi level to be thermally populated with electrons or holes.

The material's electrons seek to minimize the total energy in the material by going to low energy states; however, the Pauli exclusion principle means that they cannot all go to the lowest state. The electrons instead "fill up" the band structure starting from the bottom. The characteristic energy level up to which the electrons have filled is called the Fermi level. The position of the Fermi level with respect to the band structure is very important for electrical conduction: only electrons in energy levels near the Fermi level are free to move around, since the electrons can easily jump among the partially occupied states in that region. In contrast, the low energy states are rigidly filled with a fixed number of electrons at all times, and the high energy states are empty of electrons at all times.

In metals there are many energy levels near the Fermi level, meaning that there are many electrons available to move. This is what causes the high electronic conductivity of metals.

An important part of band theory is that there may be forbidden bands in energy: energy intervals that contain no energy levels. In insulators and semiconductors, the number of electrons happens to be just the right amount to fill a certain integer number of low energy bands, exactly to the boundary. In this case, the Fermi level falls within a band gap. Since there are no available states near the Fermi level, and the electrons are not freely movable, the electronic conductivity is very low.

## In Metals

A metal consists of a lattice of atoms, each with an outer shell of electrons that freely dissociate from their parent atoms and travel through the lattice. This is also known as a positive ionic lattice. This 'sea' of dissociable electrons allows the metal to conduct electric current. When an electrical potential difference (a voltage) is applied across the metal, the resulting electric field causes electrons to drift towards the positive terminal. The actual drift velocity of electrons is very small, in the order of magnitude of a meter per hour. However, as the electrons are densely packed in the material, the electromagnetic field is propagated through the metal at nearly the speed of light. The mechanism is similar to transfer of momentum of balls in a Newton's cradle.

Like balls in a Newton's cradle, electrons in a metal quickly transfer energy from one terminal to another, despite their own negligible movement.

Most metals have resistance. In simpler models (non quantum mechanical models) this can be explained by replacing electrons and the crystal lattice by a wave-like structure each. When the electron wave travels through the lattice the waves interfere, which causes resistance. The more regular the lattice is the less disturbance happens and thus resistance lowers. The amount of resistance is thus caused by mainly two factors. Firstly it is caused by the temperature and thus speed of vibration of the crystal lattice. The temperature causes irregularities in the lattice. Secondly the impurity of the metal is relevant as different ions cause irregularities too.

The larger the cross-sectional area of the conductor, the more electrons per unit length are available to carry the current. As a result, the resistance is lower in larger cross-section conductors. The number of scattering events encountered by an electron passing through a material is proportional to the length of the conductor. The longer the conductor, therefore, the higher the resistance. Different materials also affect the resistance.

## In Semiconductors and Insulators

In metals, the Fermi level lies in the conduction band giving rise to free conduction electrons. However, in semiconductors the position of the Fermi level is within the band gap, approximately halfway between the conduction band minimum and valence band maximum for intrinsic (undoped) semiconductors. This means that at 0 kelvin, there are no free conduction electrons, and the resistance is infinite. However, the resistance continues to decrease as the charge carrier density in the conduction band increases. In extrinsic (doped) semiconductors, dopant atoms increase the majority charge carrier concentration by donating electrons to the conduction band or producing holes in the valence band. For both types of donor or acceptor atoms, increasing dopant density reduces resistance. Hence, highly doped semiconductors behave metallically. At very high temperatures, the contribution of thermally generated carriers dominate over the contribution from dopant atoms, and the resistance decreases exponentially with temperature.

## In Ionic Liquids/Electrolytes

In electrolytes, electrical conduction happens not by band electrons or holes, but by full atomic species (ions) traveling, each carrying an electrical charge. The resistivity of ionic solutions (electrolytes) varies tremendously with concentration – while distilled water is almost an insulator, salt water is a reasonable electrical conductor. Conduction in ionic liquids is also controlled by the movement of ions, but here we are talking about molten salts rather than solvated ions. In biological membranes, currents are carried by ionic salts. Small holes in cell membranes, called ion channels, are selective to specific ions and determine the membrane resistance.

## Superconductivity

The electrical resistivity of a metallic conductor decreases gradually as temperature is lowered. In ordinary conductors, such as copper or silver, this decrease is limited by impurities and other defects. Even near absolute zero, a real sample of a normal conductor shows some resistance. In a superconductor, the resistance drops abruptly to zero when the material is cooled below its critical temperature. An electric current flowing in a loop of superconducting wire can persist indefinitely with no power source.

In 1986, researchers discovered that some cuprate-perovskite ceramic materials have much higher critical temperatures, and in 1987 one was produced with a critical temperature above 90 K (−183 °C). Such a high transition temperature is theoretically impossible for a conventional superconductor, so the researchers named these conductors *high-temperature superconductors*. Liquid nitrogen boils at 77 K, facilitating many experiments and applications that are less practical at lower temperatures. In conventional superconductors, electrons are held together in pairs by an attraction mediated by lattice phonons. The best available model of high-temperature superconductivity is still somewhat crude. There is a hypothesis that electron pairing in high-temperature superconductors is mediated by short-range spin waves known as paramagnons.

## Plasma

Lightning is an example of plasma present at Earth's surface. Typically, lightning discharges 30,000 amperes at up to 100 million volts, and emits light, radio waves, and X-rays. Plasma temperatures in lightning can approach 28,000 Kelvin (28,000 °C) (50,000 °F) and electron densities may exceed $10^{24}$ $m^{-3}$.

Plasmas are very good electrical conductors and electric potentials play an important role. The potential as it exists on average in the space between charged particles, independent of the question of how it can be measured, is called the *plasma potential*, or *space potential*. If an electrode is inserted into a plasma, its potential generally lies considerably below the plasma potential, due to what is termed a Debye sheath. The good electrical conductivity of plasmas makes their electric fields very small. This results in the important concept of *quasineutrality*, which says the density of negative charges is approximately equal to the density of positive charges over large volumes of the plasma ($n_e = \langle Z \rangle n_i$), but on the scale of the Debye length there can be charge imbalance. In the special case that *double layers* are formed, the charge separation can extend some tens of Debye lengths.

The magnitude of the potentials and electric fields must be determined by means other than simply finding the net charge density. A common example is to assume that the electrons satisfy the Boltzmann relation:

$$n_e \propto e^{e\Phi/k_B T_e}.$$

Differentiating this relation provides a means to calculate the electric field from the density:

$$\vec{E} = -\frac{k_B T_e}{e}\frac{\nabla n_e}{n_e}.$$

It is possible to produce a plasma that is not quasineutral. An electron beam, for example, has only negative charges. The density of a non-neutral plasma must generally be very low, or it must be very small. Otherwise, the repulsive electrostatic force dissipates it.

In astrophysical plasmas, Debye screening prevents electric fields from directly affecting the plasma over large distances, i.e., greater than the Debye length. However, the existence of charged particles causes the plasma to generate, and be affected by, magnetic fields. This can and does cause extremely complex behavior, such as the generation of plasma double layers, an object that separates charge over a few tens of Debye lengths. The dynamics of plasmas interacting with external and self-generated magnetic fields are studied in the academic discipline of magnetohydrodynamics.

Plasma is often called the *fourth state of matter* after solid, liquids and gases. It is distinct from these and other lower-energy states of matter. Although it is closely related to the gas phase in that it also has no definite form or volume, it differs in a number of ways, including the following:

| Property | Gas | Plasma |
|---|---|---|
| **Electrical conductivity** | Very low: air is an excellent insulator until it breaks down into plasma at electric field strengths above 30 kilovolts per centimeter. | Usually very high: for many purposes, the conductivity of a plasma may be treated as infinite. |
| **Independently acting species** | One: all gas particles behave in a similar way, influenced by gravity and by collisions with one another. | Two or three: electrons, ions, protons and neutrons can be distinguished by the sign and value of their charge so that they behave independently in many circumstances, with different bulk velocities and temperatures, allowing phenomena such as new types of waves and instabilities. |

| | | |
|---|---|---|
| **Velocity distribution** | Maxwellian: collisions usually lead to a Maxwellian velocity distribution of all gas particles, with very few relatively fast particles. | Often non-Maxwellian: collisional interactions are often weak in hot plasmas and external forcing can drive the plasma far from local equilibrium and lead to a significant population of unusually fast particles. |
| **Interactions** | Binary: two-particle collisions are the rule, three-body collisions extremely rare. | Collective: waves, or organized motion of plasma, are very important because the particles can interact at long ranges through the electric and magnetic forces. |

## Resistivity and Conductivity of Various Materials

- A conductor such as a metal has high conductivity and a low resistivity.
- An insulator like glass has low conductivity and a high resistivity.
- The conductivity of a semiconductor is generally intermediate, but varies widely under different conditions, such as exposure of the material to electric fields or specific frequencies of light, and, most important, with temperature and composition of the semiconductor material.

The degree of doping in semiconductors makes a large difference in conductivity. To a point, more doping leads to higher conductivity. The conductivity of a solution of water is highly dependent on its concentration of dissolved salts, and other chemical species that ionize in the solution. Electrical conductivity of water samples is used as an indicator of how salt-free, ion-free, or impurity-free the sample is; the purer the water, the lower the conductivity (the higher the resistivity). Conductivity measurements in water are often reported as *specific conductance*, relative to the conductivity of pure water at 25 °C. An EC meter is normally used to measure conductivity in a solution. A rough summary is as follows:

| Material | Resistivity, ρ (Ω·m) |
|---|---|
| Superconductors | 0 |
| Metals | $10^{-8}$ |
| Semiconductors | Variable |
| Electrolytes | Variable |
| Insulators | $10^{16}$ |
| Superinsulators | ∞ |

This table shows the resistivity, conductivity and temperature coefficient of various materials at 20 °C (68 °F, 293 K)

| Material | ρ (Ω·m) at 20 °C | σ (S/m) at 20 °C | Temperature coefficient ($K^{-1}$) |
|---|---|---|---|
| Carbon (graphene) | $1.00\times10^{-8}$ | $1.00\times10^{8}$ | −0.0002 |
| Silver | $1.59\times10^{-8}$ | $6.30\times10^{7}$ | 0.0038 |
| Copper | $1.68\times10^{-8}$ | $5.96\times10^{7}$ | 0.003862 |
| Annealed copper | $1.72\times10^{-8}$ | $5.80\times10^{7}$ | 0.00393 |
| Gold | $2.44\times10^{-8}$ | $4.10\times10^{7}$ | 0.0034 |
| Aluminium | $2.82\times10^{-8}$ | $3.50\times10^{7}$ | 0.0039 |

| | | | |
|---|---|---|---|
| Calcium | $3.36\times10^{-8}$ | $2.98\times10^{7}$ | 0.0041 |
| Tungsten | $5.60\times10^{-8}$ | $1.79\times10^{7}$ | 0.0045 |
| Zinc | $5.90\times10^{-8}$ | $1.69\times10^{7}$ | 0.0037 |
| Nickel | $6.99\times10^{-8}$ | $1.43\times10^{7}$ | 0.006 |
| Lithium | $9.28\times10^{-8}$ | $1.08\times10^{7}$ | 0.006 |
| Iron | $9.71\times10^{-8}$ | $1.00\times10^{7}$ | 0.005 |
| Platinum | $1.06\times10^{-7}$ | $9.43\times10^{6}$ | 0.00392 |
| Tin | $1.09\times10^{-7}$ | $9.17\times10^{6}$ | 0.0045 |
| Carbon steel (1010) | $1.43\times10^{-7}$ | $6.99\times10^{6}$ | |
| Lead | $2.20\times10^{-7}$ | $4.55\times10^{6}$ | 0.0039 |
| Titanium | $4.20\times10^{-7}$ | $2.38\times10^{6}$ | 0.0038 |
| Grain oriented electrical steel | $4.60\times10^{-7}$ | $2.17\times10^{6}$ | |
| Manganin | $4.82\times10^{-7}$ | $2.07\times10^{6}$ | 0.000002 |
| Constantan | $4.90\times10^{-7}$ | $2.04\times10^{6}$ | 0.000008 |
| Stainless steel | $6.90\times10^{-7}$ | $1.45\times10^{6}$ | 0.00094 |
| Mercury | $9.80\times10^{-7}$ | $1.02\times10^{6}$ | 0.0009 |
| Nichrome | $1.10\times10^{-6}$ | $6.7\times10^{5}$ | 0.0004 |
| GaAs | $1.00\times10^{-3}$ to $1.00\times10^{8}$ | $1.00\times10^{-8}$ to $10^{3}$ | |
| Carbon (amorphous) | $5.00\times10^{-4}$ to $8.00\times10^{-4}$ | $1.25\times10^{3}$ to $2\times10^{3}$ | −0.0005 |
| Carbon (graphite) | $2.50\times10^{-6}$ to $5.00\times10^{-6}$ $\parallel$basal plane<br>$3.00\times10^{-3}$ $\perp$basal plane | $2.00\times10^{5}$ to $3.00\times10^{5}$ $\parallel$basal plane<br>$3.30\times10^{2}$ $\perp$basal plane | |
| PEDOT:PSS | $2\times10^{-6}$ to $1\times10^{-1}$ | $1\times10^{1}$ to $4.6\times10^{5}$ | ? |
| Germanium | $4.60\times10^{-1}$ | 2.17 | −0.048 |
| Sea water | $2.00\times10^{-1}$ | 4.80 | |
| Swimming pool water | $3.33\times10^{-1}$ to $4.00\times10^{-1}$ | 0.25 to 0.30 | |
| Drinking water | $2.00\times10^{1}$ to $2.00\times10^{3}$ | $5.00\times10^{-4}$ to $5.00\times10^{-2}$ | |
| Silicon | $6.40\times10^{2}$ | $1.56\times10^{-3}$ | −0.075 |
| Wood (damp) | $1.00\times10^{3}$ to $1.00\times10^{4}$ | $10^{-4}$ to $10^{-3}$ | |
| Deionized water | $1.80\times10^{5}$ | $5.50\times10^{-6}$ | |
| Glass | $1.00\times10^{11}$ to $1.00\times10^{15}$ | $10^{-15}$ to $10^{-11}$ | ? |
| Hard rubber | $1.00\times10^{13}$ | $10^{-14}$ | ? |
| Wood (oven dry) | $1.00\times10^{14}$ to $1.00\times10^{16}$ | $10^{-16}$ to $10^{-14}$ | |
| Sulfur | $1.00\times10^{15}$ | $10^{-16}$ | ? |
| Air | $1.30\times10^{14}$ to $3.30\times10^{14}$ | $3\times10^{-15}$ to $8\times10^{-15}$ | |
| Carbon (diamond) | $1.00\times10^{12}$ | $\sim10^{-13}$ | |
| Fused quartz | $7.50\times10^{17}$ | $1.30\times10^{-18}$ | ? |
| PET | $1.00\times10^{21}$ | $10^{-21}$ | ? |
| Teflon | $1.00\times10^{23}$ to $1.00\times10^{25}$ | $10^{-25}$ to $10^{-23}$ | ? |

The effective temperature coefficient varies with temperature and purity level of the material. The 20 °C value is only an approximation when used at other temperatures. For example, the coefficient becomes lower at higher temperatures for copper, and the value 0.00427 is commonly specified at 0 °C.

The extremely low resistivity (high conductivity) of silver is characteristic of metals. George Gamow tidily summed up the nature of the metals' dealings with electrons in his science-popularizing book, *One, Two, Three...Infinity* (1947):

The metallic substances differ from all other materials by the fact that the outer shells of their atoms are bound rather loosely, and often let one of their electrons go free. Thus the interior of a metal is filled up with a large number of unattached electrons that travel aimlessly around like a crowd of displaced persons. When a metal wire is subjected to electric force applied on its opposite ends, these free electrons rush in the direction of the force, thus forming what we call an electric current.

More technically, the free electron model gives a basic description of electron flow in metals.

Wood is widely regarded as an extremely good insulator, but its resistivity is sensitively dependent on moisture content, with damp wood being a factor of at least $10^{10}$ worse insulator than oven-dry. In any case, a sufficiently high voltage – such as that in lightning strikes or some high-tension powerlines – can lead to insulation breakdown and electrocution risk even with apparently dry wood.

## Temperature Dependence

### Linear Approximation

The electrical resistivity of most materials changes with temperature. If the temperature $T$ does not vary too much, a linear approximation is typically used:

$$\rho(T) = \rho_0[1 + \alpha(T - T_0)]$$

where $\alpha$ is called the *temperature coefficient of resistivity*, $T_0$ is a fixed reference temperature (usually room temperature), and $\rho_0$ is the resistivity at temperature $T_0$. The parameter $\alpha$ is an empirical parameter fitted from measurement data. Because the linear approximation is only an approximation, $\alpha$ is different for different reference temperatures. For this reason it is usual to specify the temperature that $\alpha$ was measured at with a suffix, such as $\alpha_{15}$, and the relationship only holds in a range of temperatures around the reference. When the temperature varies over a large temperature range, the linear approximation is inadequate and a more detailed analysis and understanding should be used.

### Metals

In general, electrical resistivity of metals increases with temperature. Electron–phonon interactions can play a key role. At high temperatures, the resistance of a metal increases linearly with temperature. As the temperature of a metal is reduced, the temperature dependence of resistivity follows a power law function of temperature. Mathematically the temperature dependence of the resistivity ρ of a metal is given by the Bloch–Grüneisen formula:

$$\rho(T) = \rho(0) + A\left(\frac{T}{\Theta_R}\right)^n \int_0^{\frac{\Theta_R}{T}} \frac{x^n}{(e^x - 1)(1 - e^{-x})} dx$$

where $\rho(0)$ is the residual resistivity due to defect scattering, A is a constant that depends on the velocity of electrons at the Fermi surface, the Debye radius and the number density of electrons in the metal. $\Theta_R$ is the Debye temperature as obtained from resistivity measurements and matches very closely with the values of Debye temperature obtained from specific heat measurements. n is an integer that depends upon the nature of interaction:

1. n=5 implies that the resistance is due to scattering of electrons by phonons (as it is for simple metals)
2. n=3 implies that the resistance is due to s-d electron scattering (as is the case for transition metals)
3. n=2 implies that the resistance is due to electron–electron interaction.

If more than one source of scattering is simultaneously present, Matthiessen's Rule (first formulated by Augustus Matthiessen in the 1860s) says that the total resistance can be approximated by adding up several different terms, each with the appropriate value of *n*.

As the temperature of the metal is sufficiently reduced (so as to 'freeze' all the phonons), the resistivity usually reaches a constant value, known as the residual resistivity. This value depends not only on the type of metal, but on its purity and thermal history. The value of the residual resistivity of a metal is decided by its impurity concentration. Some materials lose all electrical resistivity at sufficiently low temperatures, due to an effect known as superconductivity.

An investigation of the low-temperature resistivity of metals was the motivation to Heike Kamerlingh Onnes's experiments that led in 1911 to discovery of superconductivity.

## Semiconductors

In general, intrinsic semiconductor resistivity decreases with increasing temperature. The electrons are bumped to the conduction energy band by thermal energy, where they flow freely, and in doing so leave behind holes in the valence band, which also flow freely. The electric resistance of a typical intrinsic (non doped) semiconductor decreases exponentially with temperature:

$$\rho = \rho_0 e^{-aT}$$

An even better approximation of the temperature dependence of the resistivity of a semiconductor is given by the Steinhart–Hart equation:

$$\frac{1}{T} = A + B\ln(\rho) + C(\ln(\rho))^3$$

where *A*, *B* and *C* are the so-called Steinhart–Hart coefficients.

This equation is used to calibrate thermistors.

Extrinsic (doped) semiconductors have a far more complicated temperature profile. As temperature increases starting from absolute zero they first decrease steeply in resistance as the carriers leave the donors or acceptors. After most of the donors or acceptors have lost their carriers, the resistance starts to increase again slightly due to the reducing mobility of carriers (much as in a metal). At higher temperatures, they behave like intrinsic semiconductors as the carriers from the donors/acceptors become insignificant compared to the thermally generated carriers.

In non-crystalline semiconductors, conduction can occur by charges quantum tunnelling from one localised site to another. This is known as variable range hopping and has the characteristic form of

$$\rho = A \exp\left(T^{-\frac{1}{n}}\right),$$

where $n = 2, 3, 4$, depending on the dimensionality of the system.

## Complex Resistivity and Conductivity

When analyzing the response of materials to alternating electric fields (dielectric spectroscopy), in applications such as electrical impedance tomography, it is convenient to replace resistivity with a complex quantity called impeditivity (in analogy to electrical impedance). Impeditivity is the sum of a real component, the resistivity, and an imaginary component, the reactivity (in analogy to reactance). The magnitude of impeditivity is the square root of sum of squares of magnitudes of resistivity and reactivity.

Conversely, in such cases the conductivity must be expressed as a complex number (or even as a matrix of complex numbers, in the case of anisotropic materials) called the *admittivity*. Admittivity is the sum of a real component called the conductivity and an imaginary component called the susceptivity.

An alternative description of the response to alternating currents uses a real (but frequency-dependent) conductivity, along with a real permittivity. The larger the conductivity is, the more quickly the alternating-current signal is absorbed by the material (i.e., the more opaque the material is).

## Tensor Equations for Anisotropic Materials

Some materials are anisotropic, meaning they have different properties in different directions. For example, a crystal of graphite consists microscopically of a stack of sheets, and current flows very easily through each sheet, but moves much less easily from one sheet to the next.

For an anisotropic material, it is not generally valid to use the scalar equations

$$J = \sigma E \rightleftharpoons E = \rho J.$$

For example, the current may not flow in exactly the same direction as the electric field. Instead, the equations are generalized to the 3D tensor form

$$\mathrm{J} = \sigma \mathrm{E} \rightleftharpoons \mathrm{E} = \rho \mathrm{J}$$

where the conductivity $\sigma$ and resistivity $\rho$ are rank-2 tensors (in other words, 3×3 matrices). The equations are compactly illustrated in component form (using index notation and the summation convention):

$$J_i = \sigma_{ij} E_j \rightleftharpoons E_i = \rho_{ij} J_j.$$

The $\sigma$ and $\rho$ tensors are inverses (in the sense of a matrix inverse). The individual components are not necessarily inverses; for example, $\sigma_{xx}$ may not be equal to $1/\rho_{xx}$.

## Resistance Versus Resistivity in Complicated Geometries

Even if the material's resistivity is known, calculating the resistance of something made from it may, in some cases, be much more complicated than the formula $R = \rho \ell / A$ above. One example is spreading resistance profiling, where the material is inhomogeneous (different resistivity in different places), and the exact paths of current flow are not obvious.

In cases like this, the formulas

$$J = \sigma E \rightleftharpoons E = \rho J$$

must be replaced with

$$\mathrm{J(r)} = \sigma(\mathrm{r})\,\mathrm{E(r)} \rightleftharpoons \mathrm{E(r)} = \rho(\mathrm{r})\,\mathrm{J(r)},$$

where E and J are now vector fields. This equation, along with the continuity equation for J and the Poisson's equation for E, form a set of partial differential equations. In special cases, an exact or approximate solution to these equations can be worked out by hand, but for very accurate answers in complex cases, computer methods like finite element analysis may be required.

## Resistivity Density Products

In some applications where the weight of an item is very important resistivity density products are more important than absolute low resistivity – it is often possible to make the conductor thicker to make up for a higher resistivity; and then a low resistivity density product material (or equivalently a high conductance to density ratio) is desirable. For example, for long distance overhead power lines, aluminium is frequently used rather than copper because it is lighter for the same conductance.

| Material | Resistivity (nΩ·m) | Density (g/cm³) | Resistivity-density | | Conductor cross-section/volume, at same conductance relative to copper |
|---|---|---|---|---|---|
| | | | (nΩ·m·g/cm³) | Relative to copper | |
| Sodium | 47.7 | 0.97 | 46 | 31% | 2.843 |
| Lithium | 92.8 | 0.53 | 49 | 33% | 5.531 |
| Calcium | 33.6 | 1.55 | 52 | 35% | 2.002 |
| Potassium | 72.0 | 0.89 | 64 | 43% | 4.291 |
| Beryllium | 35.6 | 1.85 | 66 | 44% | 2.122 |
| Aluminium | 26.50 | 2.70 | 72 | 48% | 1.5792 |
| Magnesium | 43.90 | 1.74 | 76.3 | 50.9% | 2.616 |
| Copper | 16.78 | 8.96 | 150 | 100% | 1 |
| Silver | 15.87 | 10.49 | 166 | 111% | 0.946 |
| Gold | 22.14 | 19.30 | 427 | 285% | 1.319 |
| Iron | 96.1 | 7.874 | 757 | 505% | 5.727 |

Silver, although it is the least resistive metal known, has a high density and does poorly by this measure. Calcium and the alkali metals have the best resistivity-density products, but are rarely used for conductors due to their high reactivity with water and oxygen. Aluminium is far more stable. Two other important attributes, price and toxicity, exclude the (otherwise) best choice: Beryllium. Thus, aluminium is usually the metal of choice when the weight or cost of a conductor is the driving consideration.

## Heat Capacity

Heat capacity or thermal capacity is a measurable physical quantity equal to the ratio of the heat added to (or removed from) an object to the resulting temperature change. The unit of heat capacity is joule per kelvin $\frac{J}{K}$, or kilogram metre squared per kelvin second squared $\frac{kgm^2}{Ks^2}$ in the International System of Units (SI). The dimensional form is $L^2MT^{-2}\Theta^{-1}$. Specific heat is the amount of heat needed to raise the temperature of one kilogram of mass by 1 kelvin.

Heat capacity is an extensive property of matter, meaning it is proportional to the size of the system. When expressing the same phenomenon as an intensive property, the heat capacity is divided by the amount of substance, mass, or volume, thus the quantity is independent of the size or extent of the sample. The molar heat capacity is the heat capacity per unit amount (SI unit: mole) of a pure substance and the specific heat capacity, often called simply specific heat, is the heat capacity per unit mass of a material. Nonetheless some authors use the term specific heat to refer to the ratio of the specific heat capacity of a substance at any given temperature, to the specific heat capacity of another substance at a reference temperature, much in the fashion of specific gravity. In some engineering contexts, the volumetric heat capacity is used.

Temperature reflects the average randomized kinetic energy of constituent particles of matter (i.e., atoms or molecules) relative to the centre of mass of the system, while heat is the transfer of energy across a system boundary into the body other than by work or matter transfer. Translation, rotation, and vibration of atoms represent the degrees of freedom of motion which classically contribute to the heat capacity of gases, while only vibrations are needed to describe the heat capacities of most solids, as shown by the Dulong–Petit law. Other contributions can come from magnetic and electronic degrees of freedom in solids, but these rarely make substantial contributions.

For quantum mechanical reasons, at any given temperature, some of these degrees of freedom may be unavailable, or only partially available, to store thermal energy. In such cases, the heat capacity is a fraction of the maximum. As the temperature approaches absolute zero, the heat capacity of a system approaches zero, because of loss of available degrees of freedom. Quantum theory can be used to quantitatively predict the heat capacity of simple systems.

### History

In a previous theory of heat common in the early modern period, heat was thought to be a measurement of an invisible fluid, known as the *caloric*. Bodies were capable of holding a certain amount of this fluid, hence the term *heat capacity*, named and first investigated by Scottish chemist Joseph Black in the 1750s.

Since the development of thermodynamics in the 18th and 19th centuries, scientists have abandoned the idea of a physical caloric, and instead understand heat as a manifestation of a system's internal energy. Heat is no longer considered a fluid, but rather a transfer of disordered energy. Nevertheless, at least in English, the term "heat capacity" survives. In some other languages, the term *thermal capacity* is preferred, and it is also sometimes used in English.

## Units

### Extensive Properties

In the International System of Units, heat capacity has the unit joules per kelvin (J/K). The heat capacity (symbol *C*) of a system is defined as the ratio of heat transferred to or from the system and the resulting change in temperature in the system,

$$C(T) = \frac{\delta Q}{dT},$$

where the symbol $\delta$ designates heat as a path function. If the temperature change is sufficiently small the heat capacity may be assumed to be constant:

$$C = \frac{Q}{\Delta T}.$$

Heat capacity is an extensive property, meaning it depends on the extent or size of the physical system studied. A sample containing twice the amount of substance as another sample requires the transfer of twice the amount of heat ($Q$) to achieve the same change in temperature ($\Delta T$).

### Intensive Properties

For many purposes it is more convenient to report heat capacity as an intensive property, an intrinsic characteristic of a particular substance. In practice, this is most often an expression of the property in relation to a unit of mass; in science and engineering, such properties are often prefixed with the term *specific*. International standards now recommend that specific heat capacity always refer to division by mass. The units for the specific heat capacity are $[c] = \frac{\text{J}}{\text{kg} \times \text{K}}$.

In chemistry, heat capacity is often specified relative to one mole, the unit of amount of substance, and is called the molar heat capacity. It has the unit $[C_{\text{mol}}] = \frac{\text{J}}{\text{mol} \times \text{K}}$.

For some considerations it is useful to specify the volume-specific heat capacity, commonly called volumetric heat capacity, which is the heat capacity per unit volume and has SI units $[s] = \frac{\text{J}}{\text{m}^3 \times \text{K}}$. This is used almost exclusively for liquids and solids, since for gases it may be confused with specific heat capacity *at constant volume.*

### Alternative Unit Systems

While SI units are the most widely used, some countries and industries also use other systems of measurement. One older unit of heat is the kilogram-calorie (Cal), originally defined as the energy required to raise the temperature of one kilogram of water by one degree Celsius, typically from

14.5 to 15.5 °C. The specific average heat capacity of water on this scale would therefore be exactly 1 Cal/(C°·kg). However, due to the temperature-dependence of the specific heat, a large number of different definitions of the calorie came into being. Whilst once it was very prevalent, especially its smaller cgs variant the gram-calorie (cal), defined thus the specific heat of water would be 1 cal/(K·g), in most fields the use of the calorie is now archaic.

In the United States other units of measure for heat capacity may be quoted in disciplines such as construction, civil engineering, and chemical engineering. A still common system is the English Engineering Units in which the mass reference is pound mass and the temperature is specified in degrees Fahrenheit or Rankine. One (rare) unit of heat is the pound calorie (lb-cal), defined as the amount of heat required to raise the temperature of one pound of water by one degree Celsius. On this scale the specific heat of water would be 1 lb-cal/(K·lb). More common is the British thermal unit, the standard unit of heat in the U.S. construction industry. This is defined such that the specific heat of water is 1 BTU/(F°·lb). The path integral Monte Carlo method is a numerical approach for determining the values of heat capacity, based on quantum dynamical principles. However, good approximations can be made for gases in many states using simpler methods outlined below. For many solids composed of relatively heavy atoms (atomic number > iron), at non-cryogenic temperatures, the heat capacity at room temperature approaches 3R = 24.94 joules per kelvin per mole of atoms (Dulong–Petit law, R is the gas constant). Low temperature approximations for both gases and solids at temperatures less than their characteristic Einstein temperatures or Debye temperatures can be made by the methods of Einstein and Debye discussed below. Water (liquid): CP = 4185.5 J/(kg·K) (15 °C, 101.325 kPa) Water (liquid): CVH = 74.539 J/(mol·K) (25 °C) For liquids and gases, it is important to know the pressure to which given heat capacity data refer. Most published data are given for standard pressure. However, different standard conditions for temperature and pressure have been defined by different organizations. The International Union of Pure and Applied Chemistry (IUPAC) changed its recommendation from one atmosphere to the round value 100 kPa (≈750.062 Torr).

## Measurement

It may appear that the way to measure heat capacity is to add a known amount of heat to an object, and measure the change in temperature. This works reasonably well for many solids. However, for precise measurements, and especially for gases, other aspects of measurement become critical.

The heat capacity can be affected by many of the state variables that describe the thermodynamic system under study. These include the starting and ending temperature, as well as the pressure and the volume of the system before and after heat is added. So rather than a single way to measure heat capacity, there are actually several slightly different measurements of heat capacity. The most commonly used methods for measurement are to hold the object either at constant pressure ($C_P$) or at constant volume ($C_V$). Gases and liquids are typically also measured at constant volume. Measurements under constant pressure produce larger values than those at constant volume because the constant pressure values also include heat energy that is used to do work to expand the substance against the constant pressure as its temperature increases. This difference is particularly notable in gases where values under constant pressure are typically 30% to 66.7% greater than those at constant volume. Hence the heat capacity ratio of gases is typically between 1.3 and 1.67.

The specific heat capacities of substances comprising molecules (as distinct from monatomic gases) are not fixed constants and vary somewhat depending on temperature. Accordingly, the temperature at which the measurement is made is usually also specified. Examples of two common ways to cite the specific heat of a substance are as follows:

- Water (liquid): $C_P$ = 4185.5 J/(kg·K) (15 °C, 101.325 kPa)
- Water (liquid): $C_V H$ = 74.539 J/(mol·K) (25 °C)

For liquids and gases, it is important to know the pressure to which given heat capacity data refer. Most published data are given for standard pressure. However, quite different standard conditions for temperature and pressure have been defined by different organizations. The International Union of Pure and Applied Chemistry (IUPAC) changed its recommendation from one atmosphere to the round value 100 kPa (≈750.062 Torr).

## Calculation from First Principles

The path integral Monte Carlo method is a numerical approach for determining the values of heat capacity, based on quantum dynamical principles. However, good approximations can be made for gases in many states using simpler methods outlined below. For many solids composed of relatively heavy atoms (atomic number > iron), at non-cryogenic temperatures, the heat capacity at room temperature approaches $3R$ = 24.94 joules per kelvin per mole of atoms (Dulong–Petit law, $R$ is the gas constant). Low temperature approximations for both gases and solids at temperatures less than their characteristic Einstein temperatures or Debye temperatures can be made by the methods of Einstein and Debye discussed below.

## Thermodynamic Relations and Definition of Heat Capacity

The internal energy of a closed system changes either by adding heat to the system or by the system performing work. Written mathematically we have

$$\Delta e_{\text{system}} = e_{\text{in}} - e_{\text{out}}$$

or

$$\mathrm{d}U = \delta Q + \delta W.$$

For work as a result of an increase of the system volume we may write,

$$\mathrm{d}U = \delta Q - P\mathrm{d}V.$$

If the heat is added at constant volume, then the second term of this relation vanishes and one readily obtains

$$\left(\frac{\partial U}{\partial T}\right)_{\text{V}} = \left(\frac{\partial Q}{\partial T}\right)_{\text{V}} = C_{\text{V}}.$$

This defines the *heat capacity at constant volume*, $C_V$, which is also related to changes in internal energy. Another useful quantity is the *heat capacity at constant pressure*, $C_P$. This quantity refers to the change in the *enthalpy* of the system, which is given by

$$H = U + PV.$$

A small change in the enthalpy can be expressed as

$$dH = \delta Q + V\,dP,$$

and therefore, at constant pressure, we have

$$\left(\frac{\partial H}{\partial T}\right)_P = \left(\frac{\partial Q}{\partial T}\right)_P = C_P.$$

These two equations:

$$\left(\frac{\partial U}{\partial T}\right)_V = \left(\frac{\partial Q}{\partial T}\right)_V = C_V.$$

$$\left(\frac{\partial H}{\partial T}\right)_P = \left(\frac{\partial Q}{\partial T}\right)_P = C_P.$$

are property relations and are therefore independent of the type of process. In other words, they are valid for any substance going through any process. Both the internal energy and enthalpy of a substance can change with the transfer of energy in many forms i.e., heat.

## Relation between Heat Capacities

Measuring the heat capacity, sometimes referred to as specific heat, at constant volume can be prohibitively difficult for liquids and solids. That is, small temperature changes typically require large pressures to maintain a liquid or solid at constant volume implying the containing vessel must be nearly rigid or at least very strong. Instead it is easier to measure the heat capacity at constant pressure (allowing the material to expand or contract freely) and solve for the heat capacity at constant volume using mathematical relationships derived from the basic thermodynamic laws. Starting from the fundamental thermodynamic relation one can show

$$C_P - C_V = T\left(\frac{\partial P}{\partial T}\right)_{V,n}\left(\frac{\partial V}{\partial T}\right)_{P,n}$$

where the partial derivatives are taken at constant volume and constant number of particles, and constant pressure and constant number of particles, respectively.

This can also be rewritten

$$C_P - C_V = VT\frac{\alpha^2}{\beta_T}$$

where

$\alpha$ is the coefficient of thermal expansion,

$\beta_T$ is the isothermal compressibility.

The heat capacity ratio or adiabatic index is the ratio of the heat capacity at constant pressure to heat capacity at constant volume. It is sometimes also known as the isentropic expansion factor.

## Ideal Gas

For an ideal gas, evaluating the partial derivatives above according to the equation of state where R is the gas constant for an ideal gas

$$PV = nRT$$

$$C_P - C_V = T\left(\frac{\partial P}{\partial T}\right)_{V,n}\left(\frac{\partial V}{\partial T}\right)_{P,n}$$

$$P = \frac{nRT}{V} \Rightarrow \left(\frac{\partial P}{\partial T}\right)_{V,n} = \frac{nR}{V}$$

$$V = \frac{nRT}{P} \Rightarrow \left(\frac{\partial V}{\partial T}\right)_{P,n} = \frac{nR}{P}$$

substituting

$$T\left(\frac{\partial P}{\partial T}\right)_{V,n}\left(\frac{\partial V}{\partial T}\right)_{P,n} = T\left(\frac{nR}{V}\right)\left(\frac{nR}{P}\right) = \left(\frac{nRT}{V}\right)\left(\frac{nR}{P}\right) = P\left(\frac{nR}{P}\right) = nR$$

this equation reduces simply to Mayer's relation,

$$C_{P,m} - C_{V,m} = R$$

## Specific Heat Capacity

The specific heat capacity of a material on a per mass basis is

$$c = \frac{\partial C}{\partial m},$$

which in the absence of phase transitions is equivalent to

$$c = E_m = \frac{C}{m} = \frac{C}{\rho V},$$

where

$C$ is the heat capacity of a body made of the material in question,

$m$ is the mass of the body,

$V$ is the volume of the body, and

$\rho = \frac{m}{V}$ is the density of the material.

For gases, and also for other materials under high pressures, there is need to distinguish between different boundary conditions for the processes under consideration (since values differ significantly between different conditions). Typical processes for which a heat capacity may be defined include isobaric (constant pressure, $dP = 0$) or isochoric (constant volume, $dV = 0$) processes. The corresponding specific heat capacities are expressed as

$$c_P = \left( \frac{\partial C}{\partial m} \right)_P ,$$

$$c_V = \left( \frac{\partial C}{\partial m} \right)_V .$$

From the results of the last section, dividing through by the mass gives the relation

$$c_P - c_V = \frac{\alpha^2 T}{\rho \beta_T} .$$

A related parameter to $c$ is $CV^{-1}$, the volumetric heat capacity. In engineering practice, $c_V$ for solids or liquids often signifies a volumetric heat capacity, rather than a constant-volume one. In such cases, the mass-specific heat capacity (specific heat) is often explicitly written with the subscript $m$, as $c_m$. Of course, from the above relationships, for solids one writes

$$c_m = \frac{C}{m} = \frac{c_{\text{volumetric}}}{\rho} .$$

For pure homogeneous chemical compounds with established molecular or molar mass, or a molar quantity, heat capacity as an intensive property can be expressed on a per mole basis instead of a per mass basis by the following equations analogous to the per mass equations:

$C_{P,m} = \left( \frac{\partial C}{\partial n} \right)_P$ = molar heat capacity at constant pressure

$C_{V,m} = \left( \frac{\partial C}{\partial n} \right)_V$ = molar heat capacity at constant volume

where $n$ is the number of moles in the body or thermodynamic system. One may refer to such a *per mole* quantity as molar heat capacity to distinguish it from specific heat capacity on a per mass basis.

## Polytropic Heat Capacity

The polytropic heat capacity is calculated at processes if all the thermodynamic properties (pressure, volume, temperature) change,

$C_{i,m} = \left(\frac{\partial C}{\partial n}\right)$ = molar heat capacity at polytropic process.

The most important polytropic processes run between the adiabatic and the isotherm functions, the polytropic index is between 1 and the adiabatic exponent (γ or κ).

## Dimensionless Heat Capacity

The dimensionless heat capacity of a material is

$$C^* = \frac{C}{nR} = \frac{C}{Nk}$$

where

$C$ is the heat capacity of a body made of the material in question (J/K)

$n$ is the amount of substance in the body (mol)

$R$ is the gas constant (J/(K·mol))

$N$ is the number of molecules in the body. (dimensionless)

$k$ is Boltzmann's constant (J/(K·molecule))

In the ideal gas, dimensionless heat capacity $C^*$ is expressed as $\hat{c}$, and is related there directly to half the number of degrees of freedom per particle. This holds true for quadratic degrees of freedom, a consequence of the equipartition theorem.

More generally, the dimensionless heat capacity relates the logarithmic increase in temperature to the increase in the dimensionless entropy per particle $S^* = S / Nk$, measured in nats.

$$C^* = \frac{\mathrm{d}S^*}{\mathrm{d}\ln T}$$

Alternatively, using base 2 logarithms, $C^*$ relates the base-2 logarithmic increase in temperature to the increase in the dimensionless entropy measured in bits.

## Heat Capacity at Absolute Zero

From the definition of entropy

$$T\,\mathrm{d}S = \delta Q$$

the absolute entropy can be calculated by integrating from zero kelvins temperature to the final temperature $T_f$

$$S(T_f) = \int_{T=0}^{T_f} \frac{\delta Q}{T} = \int_0^{T_f} \frac{\delta Q}{dT}\frac{dT}{T} = \int_0^{T_f} C(T)\frac{dT}{T}.$$

The heat capacity must be zero at zero temperature in order for the above integral not to yield an infinite absolute entropy, which would violate the third law of thermodynamics. One of the strengths of the Debye model is that (unlike the preceding Einstein model) it predicts the proper mathematical form of the approach of heat capacity toward zero, as absolute zero temperature is approached.

## Negative Heat Capacity (Stars)

Most physical systems exhibit a positive heat capacity. However, even though it can seem paradoxical at first, there are some systems for which the heat capacity is *negative*. These are inhomogeneous systems which do not meet the strict definition of thermodynamic equilibrium. They include gravitating objects such as stars, galaxies; and also sometimes some nano-scale clusters of a few tens of atoms, close to a phase transition. A negative heat capacity can result in a negative temperature.

According to the virial theorem, for a self-gravitating body like a star or an interstellar gas cloud, the average potential energy $U_{Pot}$ and the average kinetic energy $U_{Kin}$ are locked together in the relation

$$U_{Pot} = -2U_{Kin},$$

The total energy $U$ (= $U_{Pot} + U_{Kin}$) therefore obeys

$$U = -U_{Kin},$$

If the system loses energy, for example by radiating energy away into space, the average kinetic energy actually increases. If a temperature is defined by the average kinetic energy, then the system therefore can be said to have a negative heat capacity.

A more extreme version of this occurs with black holes. According to black hole thermodynamics, the more mass and energy a black hole absorbs, the colder it becomes. In contrast, if it is a net emitter of energy, through Hawking radiation, it will become hotter and hotter until it boils away.

## Theory

## Factors that Affect Specific Heat Capacity

For any given substance, the heat capacity of a body is directly proportional to the amount of substance it contains (measured in terms of mass or moles or volume). Doubling the amount of substance in a body doubles its heat capacity, etc.

However, when this effect has been corrected for, by dividing the heat capacity by the quantity of substance in a body, the resulting specific heat capacity is a function of the structure of the substance itself. In particular, it depends on the number of degrees of freedom that are available to the particles in the substance; each independent degree of freedom allows the particles to store thermal energy. The translational kinetic energy of substance particles which manifests as *temperature change* is only one of the many possible degrees of freedom, and thus the larger the number of degrees of freedom available to the particles of a substance *other* than translational kinetic energy, the larger will be the specific heat capacity for the substance. For example, rotational kinetic energy of gas molecules stores heat energy in a way that increases heat capacity, since this energy does not contribute to temperature.

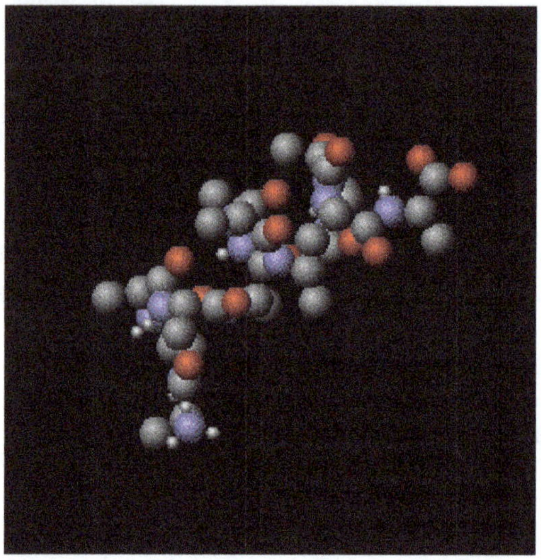

Molecules undergo many characteristic internal vibrations. Potential energy stored in these internal degrees of freedom contributes to a sample's energy content, but not to its temperature. More internal degrees of freedom tend to increase a substance's specific heat capacity, so long as temperatures are high enough to overcome quantum effects.

In addition, quantum effects require that whenever energy be stored in any mechanism associated with a bound system which confers a degree of freedom, it must be stored in certain minimal-sized deposits (quanta) of energy, or else not stored at all. Such effects limit the full ability of some degrees of freedom to store energy when their lowest energy storage quantum amount is not easily supplied at the average energy of particles at a given temperature. In general, for this reason, specific heat capacities tend to fall at lower temperatures where the average thermal energy available to each particle degree of freedom is smaller, and thermal energy storage begins to be limited by these quantum effects. Due to this process, as temperature falls toward absolute zero, so also does heat capacity.

## Degrees of Freedom

Molecules are quite different from the monatomic gases like helium and argon. With monatomic gases, thermal energy comprises only translational motions. Translational motions are ordinary, whole-body movements in 3D space whereby particles move about and exchange energy in collisions—like rubber balls in a vigorously shaken container. These simple movements in the three dimensions of space mean individual atoms have three translational degrees of freedom. A degree of freedom is any form of energy in which heat transferred into an object can be stored. This can be in translational kinetic energy, rotational kinetic energy, or other forms such as potential energy in vibrational modes. Only three translational degrees of freedom (corresponding to the three independent directions in space) are available for any individual atom, whether it is free, as a monatomic molecule, or bound into a polyatomic molecule.

As to rotation about an atom's axis (again, whether the atom is bound or free), its energy of rotation is proportional to the moment of inertia for the atom, which is extremely small compared to moments of inertia of collections of atoms. This is because almost all of the mass of a single atom is concentrated in its nucleus, which has a radius too small to give a significant moment of inertia. In contrast, the *spacing* of quantum energy levels for a rotating object is inversely proportional to its moment of inertia, and so this spacing becomes very large for objects with very small moments of inertia. For these reasons, the contribution from rotation of atoms on their axes is essentially zero in monatomic gases, because the energy spacing of the associated quantum levels is too large for significant thermal energy to be stored in rotation of systems with such small moments of inertia. For similar reasons, axial rotation around bonds joining atoms in diatomic gases (or along the linear axis in a linear molecule of any length) can also be neglected as a possible "degree of freedom" as well, since such rotation is similar to rotation of monatomic atoms, and so occurs about an axis with a moment of inertia too small to be able to store significant heat energy.

In polyatomic molecules, other rotational modes may become active, due to the much higher moments of inertia about certain axes which do not coincide with the linear axis of a linear molecule. These modes take the place of some translational degrees of freedom for individual atoms, since the atoms are moving in 3-D space, as the molecule rotates. The narrowing of quantum mechanically determined energy spacing between rotational states results from situations where atoms are rotating around an axis that does not connect them, and thus form an assembly that has a large moment of inertia. This small difference between energy states allows the kinetic energy of this type of rotational motion to store heat energy at ambient temperatures. Furthermore, internal vibrational degrees of freedom also may become active (these are also a type of translation, as seen from the view of each atom). In summary, molecules are complex objects with a population of atoms that may move about within the molecule in a number of different ways, and each of these ways of moving is capable of storing energy if the temperature is sufficient.

The heat capacity of molecular substances (on a "per-atom" or atom-molar, basis) does not exceed the heat capacity of monatomic gases, unless vibrational modes are brought into play. The reason for this is that vibrational modes allow energy to be stored as potential energy in inter-atomic bonds in a molecule, which are not available to atoms in monatomic gases. Up to about twice as much energy (on a per-atom basis) per unit of temperature increase can be stored in a solid as in a monatomic gas, by this mechanism of storing energy in the potentials of interatomic bonds. This gives many solids about twice the atom-molar heat capacity at room temperature of monatomic gases.

However, quantum effects heavily affect the actual ratio at lower temperatures (i.e., much lower than the melting temperature of the solid), especially in solids with light and tightly bound atoms (e.g., beryllium metal or diamond). Polyatomic gases store intermediate amounts of energy, giving them a "per-atom" heat capacity that is between that of monatomic gases ($^3/_2\, R$ per mole of atoms, where $R$ is the ideal gas constant), and the maximum of fully excited warmer solids ($3\, R$ per mole of atoms). For gases, heat capacity never falls below the minimum of $^3/_2\, R$ per mole (of molecules), since the kinetic energy of gas molecules is always available to store at least this much thermal energy. However, at cryogenic temperatures in solids, heat capacity falls toward zero, as temperature approaches absolute zero.

## Example of Temperature-dependent Specific Heat Capacity, in a Diatomic gas

To illustrate the role of various degrees of freedom in storing heat, we may consider nitrogen, a diatomic molecule that has five active degrees of freedom at room temperature: the three comprising translational motions plus two rotational degrees of freedom internally. Although the constant-volume molar heat capacity of nitrogen at this temperature is five-thirds that of monatomic gases, on a per-mole of atoms basis, it is five-sixths that of a monatomic gas. The reason for this is the loss of a degree of freedom due to the bond when it does not allow storage of thermal energy. Two separate nitrogen atoms would have a total of six degrees of freedom—the three translational degrees of freedom of each atom. When the atoms are bonded the molecule will still only have three translational degrees of freedom, as the two atoms in the molecule move as one. However, the molecule cannot be treated as a point object, and the moment of inertia has increased sufficiently about two axes to allow two rotational degrees of freedom to be active at room temperature to give five degrees of freedom. The moment of inertia about the third axis remains small, as this is the axis passing through the centres of the two atoms, and so is similar to the small moment of inertia for atoms of a monatomic gas. Thus, this degree of freedom does not act to store heat, and does not contribute to the heat capacity of nitrogen. The heat capacity *per atom* for nitrogen (5/2 R per mole molecules = 5/4 R per mole atoms) is therefore less than for a monatomic gas (3/2 R per mole molecules or atoms), so long as the temperature remains low enough that no vibrational degrees of freedom are activated.

At higher temperatures, however, nitrogen gas gains one more degree of internal freedom, as the molecule is excited into higher vibrational modes that store thermal energy. A vibrational degree of freedom contributes a heat capacity of 1/2 R each for kinetic and potential energy, for a total of R. Now the bond is contributing heat capacity, and (because of storage of energy in potential energy) is contributing more than if the atoms were not bonded. With full thermal excitation of bond vibration, the heat capacity per volume, or per mole of gas *molecules* approaches seven-thirds that of monatomic gases. Significantly, this is seven-sixths of the monatomic gas value on a mole-of-atoms basis, so this is now a *higher* heat capacity *per atom* than the monatomic figure, because the vibrational mode enables for diatomic gases allows an extra degree of *potential energy* freedom per pair of atoms, which monatomic gases cannot possess.

However, even at these large temperatures where gaseous nitrogen is able to store $7/6^{\text{ths}}$ of the energy *per atom* of a monatomic gas (making it more efficient at storing energy on an atomic basis), it still only stores $7/12^{\text{ths}}$ of the maximal per-atom heat capacity of a *solid,* meaning it is not nearly as efficient at storing thermal energy on an atomic basis, as solid substances can be. This is typical of gases, and results because many of the potential bonds which might be storing potential energy in gaseous nitrogen (as opposed to solid nitrogen) are lacking, because only one of the spatial dimensions for each nitrogen atom offers a bond into which potential energy can be stored without increasing the kinetic energy of the atom. In general, solids are most efficient, on an atomic basis, at storing thermal energy (that is, they have the highest per-atom or per-mole-of-atoms heat capacity).

## Per Mole of Different Units

### Per Mole of Molecules

When the specific heat capacity, *c*, of a material is measured (lowercase *c* means the unit quantity is in terms of mass), different values arise because different substances have different molar masses (essentially, the weight of the individual atoms or molecules). In solids, thermal energy arises due to the number of atoms that are vibrating. “Molar” heat capacity *per mole of molecules*, for both gases and solids, offer figures which are arbitrarily large, since molecules may be arbitrarily large. Such heat capacities are thus not intensive quantities for this reason, since the quantity of mass being considered can be increased without limit.

## Per Mole of Atoms

Conversely, for *molecular-based* substances (which also absorb heat into their internal degrees of freedom), massive, complex molecules with high atomic count—like octane—can store a great deal of energy per mole and yet are quite unremarkable on a mass basis, or on a per-atom basis. This is because, in fully excited systems, heat is stored independently by each atom in a substance, not primarily by the bulk motion of molecules.

Thus, it is the heat capacity per-mole-of-atoms, not per-mole-of-molecules, which is the intensive quantity, and which comes closest to being a constant for all substances at high temperatures. This relationship was noticed empirically in 1819, and is called the Dulong–Petit law, after its two discoverers. Historically, the fact that specific heat capacities are approximately equal when corrected by the presumed weight of the atoms of solids, was an important piece of data in favor of the atomic theory of matter.

Because of the connection of heat capacity to the number of atoms, some care should be taken to specify a mole-of-molecules basis vs. a mole-of-atoms basis, when comparing specific heat capacities of molecular solids and gases. Ideal gases have the same numbers of molecules per volume, so increasing molecular complexity adds heat capacity on a per-volume and per-mole-of-molecules basis, but may lower or raise heat capacity on a per-atom basis, depending on whether the temperature is sufficient to store energy as atomic vibration.

In solids, the quantitative limit of heat capacity in general is about 3 $R$ per mole of atoms, where $R$ is the ideal gas constant. This 3 $R$ value is about 24.9 J/mole.K. Six degrees of freedom (three kinetic and three potential) are available to each atom. Each of these six contributes $\frac{1}{2}$R specific heat capacity per mole of atoms. This limit of 3 $R$ per mole specific heat capacity is approached at room temperature for most solids, with significant departures at this temperature only for solids composed of the lightest atoms which are bound very strongly, such as beryllium (where the value is only of 66% of 3 $R$), or diamond (where it is only 24% of 3 $R$). These large departures are due to quantum effects which prevent full distribution of heat into all vibrational modes, when the energy difference between vibrational quantum states is very large compared to the average energy available to each atom from the ambient temperature.

For monatomic gases, the specific heat is only half of 3 $R$ per mole, i.e. ($\frac{3}{2}$R per mole) due to loss of all potential energy degrees of freedom in these gases. For polyatomic gases, the heat capacity will be intermediate between these values on a per-mole-of-atoms basis, and (for heat-stable molecules) would approach the limit of 3 $R$ per mole of atoms, for gases composed of complex molecules, and at higher temperatures at which all vibrational modes accept excitational energy. This is because very large and complex gas molecules may be thought of as relatively large blocks of solid matter which have lost only a relatively small fraction of degrees of freedom, as compared to a fully integrated solid.

For a list of heat capacities per atom-mole of various substances, in terms of R, see the last column of the table of heat capacities below.

## Corollaries of these Considerations for Solids (Volume-specific Heat Capacity)

Since the bulk density of a solid chemical element is strongly related to its molar mass (usually about 3 $R$ per mole, as noted above), there exists a noticeable inverse correlation between a solid's density and its specific heat capacity on a per-mass basis. This is due to a very approximate tendency of atoms of most elements to be about the same size (and constancy of mole-specific heat capacity) resulting in a good correlation between the *volume* of any given solid chemical element and its total heat capacity. Another way of stating this, is that the volume-specific heat capacity (volumetric heat capacity) of solid elements is roughly a constant. The molar volume of solid elements is very roughly constant, and (even more reliably) so also is the molar heat capacity for most solid substances. These two factors determine the volumetric heat capacity, which as a bulk property may be striking in consistency. For example, the element uranium is a metal which has a density almost 36 times that of the metal lithium, but uranium's specific heat capacity on a volumetric basis (i.e. per given volume of metal) is only 18% larger than lithium's.

Since the volume-specific corollary of the Dulong–Petit specific heat capacity relationship requires that atoms of all elements take up (on average) the same volume in solids, there are many departures from it, with most of these due to variations in atomic size. For instance, arsenic, which is only 14.5% less dense than antimony, has nearly 59% more specific heat capacity on a mass basis. In other words; even though an ingot of arsenic is only about 17% larger than an antimony one of the same mass, it absorbs about 59% more heat for a given temperature rise. The heat capacity ratios of the two substances closely follows the ratios of their molar volumes (the ratios of numbers of atoms in the same volume of each substance); the departure from the correlation to simple volumes in this case is due to lighter arsenic atoms being significantly more closely packed than antimony atoms, instead of similar size. In other words, similar-sized atoms would cause a mole of arsenic to be 63% larger than a mole of antimony, with a correspondingly lower density, allowing its volume to more closely mirror its heat capacity behavior.

## Other Factors

## Hydrogen Bonds

Hydrogen-containing polar molecules like ethanol, ammonia, and water have powerful, intermolecular hydrogen bonds when in their liquid phase. These bonds provide another place where heat may be stored as potential energy of vibration, even at comparatively low temperatures. Hydrogen bonds account for the fact that liquid water stores nearly the theoretical limit of 3 $R$ per mole of atoms, even at relatively low temperatures (i.e. near the freezing point of water).

## Impurities

In the case of alloys, there are several conditions in which small impurity concentrations can greatly affect the specific heat. Alloys may exhibit marked difference in behaviour even in the case of small amounts of impurities being one element of the alloy; for example impurities in semiconducting ferromagnetic alloys may lead to quite different specific heat properties.

## The Simple Case of the Monatomic Gas

In the case of a monatomic gas such as helium under constant volume, if it is assumed that no electronic or nuclear quantum excitations occur, each atom in the gas has only 3 degrees of freedom, all of a translational type. No energy dependence is associated with the degrees of freedom which define the position of the atoms. While, in fact, the degrees of freedom corresponding to the momenta of the atoms are quadratic, and thus contribute to the heat capacity. There are $N$ atoms, each of which has 3 components of momentum, which leads to $3N$ total degrees of freedom. This gives:

$$C_V = \left(\frac{\partial U}{\partial T}\right)_V = \frac{3}{2} N k_B = \frac{3}{2} nR$$

$$C_{V,m} = \frac{C_V}{n} = \frac{3}{2} R$$

where

$C_V$ is the heat capacity at constant volume of the gas

$C_{V,m}$ is the *molar heat capacity* at constant volume of the gas

$N$ is the total number of atoms present in the container

$n$ is the number of moles of atoms present in the container ($n$ is the ratio of $N$ and Avogadro's number)

$R$ is the ideal gas constant, (8.3144621 J/(mol·K). $R$ is equal to the product of Boltzmann's constant $k_B$ and Avogadro's number

The following table shows experimental molar constant volume heat capacity measurements taken for each noble monatomic gas (at 1 atm and 25 °C):

| Monatomic gas | $C_{V,m}$ (J/(mol·K)) | $C_{V,m}/R$ |
|---|---|---|
| He | 12.5 | 1.50 |
| Ne | 12.5 | 1.50 |
| Ar | 12.5 | 1.50 |
| Kr | 12.5 | 1.50 |
| Xe | 12.5 | 1.50 |

It is apparent from the table that the experimental heat capacities of the monatomic noble gases agrees with this simple application of statistical mechanics to a very high degree.

The molar heat capacity of a monatomic gas at constant pressure is then

$$C_{p,m} = C_{V,m} + R = \frac{5}{2} R$$

## Diatomic Gas

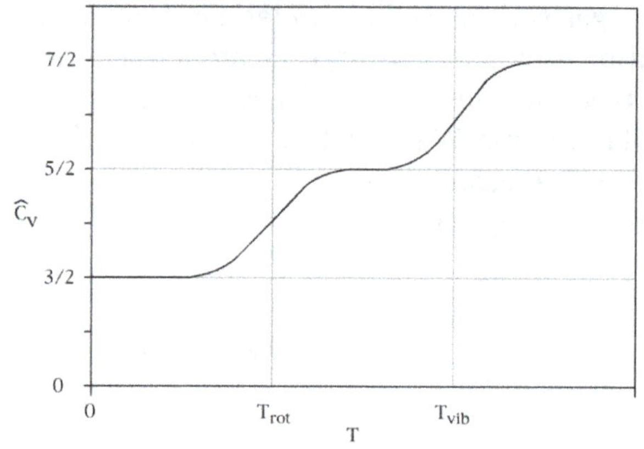


Constant volume specific heat capacity of a diatomic gas (idealised). As temperature increases, heat capacity goes from 3/2 R (translation contribution only), to 5/2 R (translation plus rotation), finally to a maximum of 7/2 R (translation + rotation + vibration).

In the somewhat more complex case of an ideal gas of diatomic molecules, the presence of internal degrees of freedom are apparent. In addition to the three translational degrees of freedom, there are rotational and vibrational degrees of freedom. In general, the number of degrees of freedom, $f$, in a molecule with $n_a$ atoms is $3n_a$:

$$f = 3n_a$$

Mathematically, there are a total of three rotational degrees of freedom, one corresponding to rotation about each of the axes of three-dimensional space. However, in practice only the existence of two degrees of rotational freedom for linear molecules will be considered. This approximation is valid because the moment of inertia about the internuclear axis is vanishingly small with respect to other moments of inertia in the molecule (this is due to the very small rotational moments of single atoms, due to the concentration of almost all their mass at their centers; compare also the extremely small radii of the atomic nuclei compared to the distance between them in a diatomic molecule). Quantum mechanically, it can be shown that the interval between successive rotational energy eigenstates is inversely proportional to the moment of inertia about that axis. Because the moment of inertia about the internuclear axis is vanishingly small relative to the other two rotational axes, the energy spacing can be considered so high that no excitations of the rotational state can occur unless the temperature is extremely high. It is easy to calculate the expected number of vibrational degrees of freedom (or vibrational modes). There are three degrees of translational freedom, and two degrees of rotational freedom, therefore

$$f_{\mathrm{vib}} = f - f_{\mathrm{trans}} - f_{\mathrm{rot}} = 6 - 3 - 2 = 1$$

Each rotational and translational degree of freedom will contribute $R/2$ in the total molar heat capacity of the gas. Each vibrational mode will contribute to the total molar heat capacity, however. This is because for each vibrational mode, there is a potential and kinetic energy component. Both the potential and kinetic components will contribute $R/2$ to the total molar heat capacity of the gas. Therefore, a diatomic molecule would be expected to have a molar constant-volume heat capacity of

$$C_{V,m} = \frac{3R}{2} + R + R = \frac{7R}{2} = 3.5R$$

where the terms originate from the translational, rotational, and vibrational degrees of freedom, respectively.

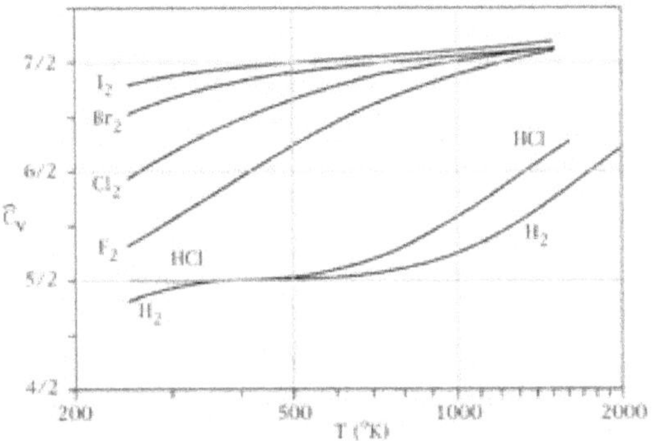


Constant volume specific heat capacity of diatomic gases (real gases) between about 200 K and 2000 K. This temperature range is not large enough to include both quantum transitions in all gases. Instead, at 200 K, all but hydrogen are fully rotationally excited, so all have at least 5/2 R heat capacity. (Hydrogen is already below 5/2, but it will require cryogenic conditions for even H2 to fall to 3/2 R). Further, only the heavier gases fully reach 7/2 R at the highest temperature, due to the relatively small vibrational energy spacing of these molecules. HCl and H2 begin to make the transition above 500 K, but have not achieved it by 1000 K, since their vibrational energy level spacing is too wide to fully participate in heat capacity, even at this temperature.

The following is a table of some molar constant-volume heat capacities of various diatomic gases at standard temperature (25 °C = 298 K)

| **Diatomic gas** | **$C_{V,m}$ (J/(mol·K))** | **$C_{V,m}$ / R** |
|---|---|---|
| $H_2$ | 20.18 | 2.427 |
| CO | 20.2 | 2.43 |
| $N_2$ | 19.9 | 2.39 |
| $Cl_2$ | 24.1 | 3.06 |
| $Br_2$ (vapour) | 28.2 | 3.39 |

From the above table, clearly there is a problem with the above theory. All of the diatomics examined have heat capacities that are lower than those predicted by the equipartition theorem, except $Br_2$. However, as the atoms composing the molecules become heavier, the heat capacities move closer to their expected values. One of the reasons for this phenomenon is the quantization of vibrational, and to a lesser extent, rotational states. In fact, if it is assumed that the molecules remain in their lowest energy vibrational state because the inter-level energy spacings for vibration-energies are large, the predicted molar constant volume heat capacity for a diatomic molecule becomes just that from the contributions of translation and rotation:

$$C_{V,m} = \frac{3R}{2} + R = \frac{5R}{2} = 2.5R$$

which is a fairly close approximation of the heat capacities of the lighter molecules in the above table. If the quantum harmonic oscillator approximation is made, it turns out that the quantum vibrational energy level spacings are actually inversely proportional to the square root of the reduced mass of the atoms composing the diatomic molecule. Therefore, in the case of the heavier diatomic molecules such as chlorine or bromine, the quantum vibrational energy level spacings become finer, which allows more excitations into higher vibrational levels at lower temperatures. This limit for storing heat capacity in vibrational modes, as discussed above, becomes 7R/2 = 3.5 R per mole of gas molecules, which is fairly consistent with the measured value for $Br_2$ at room temperature. As temperatures rise, all diatomic gases approach this value.

## General Gas Phase

The specific heat of the gas is best conceptualized in terms of the degrees of freedom of an individual molecule. The different degrees of freedom correspond to the different ways in which the molecule may store energy. The molecule may store energy in its translational motion according to the formula:

$$E = \frac{1}{2} m \left( v_x^2 + v_y^2 + v_z^2 \right)$$

where $m$ is the mass of the molecule and $[v_x, v_y, v_z]$ is velocity of the center of mass of the molecule. Each direction of motion constitutes a degree of freedom, thus there are three translational degrees of freedom.

In addition, a molecule may have rotational motion. The kinetic energy of rotational motion is generally expressed as

$$E = \frac{1}{2} \left( I_1 \omega_1^2 + I_2 \omega_2^2 + I_3 \omega_3^2 \right)$$

where $I$ is the moment of inertia tensor of the molecule, and $[\omega_1, \omega_2, \omega_3]$ is the angular velocity pseudo-vector (in a coordinate system aligned with the principal axes of the molecule). In general, then, there will be three additional degrees of freedom corresponding to the rotational motion of the molecule, (For linear molecules one of the inertia tensor terms vanishes and there are only two rotational degrees of freedom). The degrees of freedom corresponding to translations and rotations are called the rigid degrees of freedom, since they do not involve any deformation of the molecule.

The motions of the atoms in a molecule which are not part of its gross translational motion or rotation may be classified as vibrational motions. It can be shown that if there are $n$ atoms in the molecule, there will be as many as $v = 3n - 3 - n_r$ vibrational degrees of freedom, where $n_r$ is the number of rotational degrees of freedom. A vibrational degree of freedom corresponds to a specific way in which all the atoms of a molecule can vibrate. The actual number of possible vibrations may be less than this maximal one, due to various symmetries.

For example, triatomic nitrous oxide $N_2O$ will have only 2 degrees of rotational freedom (since it is a linear molecule) and contains n=3 atoms: thus the number of possible vibrational degrees of freedom will be $v = (3 \cdot 3) - 3 - 2 = 4$. There are four ways or "modes" in which the three atoms can vibrate, corresponding to *1)* A mode in which an atom at each end of the molecule moves away from, or towards, the center atom at the same time, *2)* a mode in which either end atom moves asynchronously with regard to the other two, and *3)* and *4)* two modes in which the molecule bends out of line, from the center, in the two possible planar directions that are orthogonal to its axis. Each vibrational degree of freedom confers TWO total degrees of freedom, since vibrational energy mode partitions into 1 kinetic and 1 potential mode. This would give nitrous oxide 3 translational, 2 rotational, and 4 vibrational modes (but these last giving 8 vibrational degrees of freedom), for storing energy. This is a total of $f = 3 + 2 + 8 = 13$ total energy-storing degrees of freedom, for $N_2O$.

For a bent molecule like water $H_2O$, a similar calculation gives $9 - 3 - 3 = 3$ modes of vibration, and 3 (translational) + 3 (rotational) + 6 (vibrational) = 12 degrees of freedom.

## The Storage of Energy into Degrees of Freedom

If the molecule could be entirely described using classical mechanics, then the theorem of equipartition of energy could be used to predict that each degree of freedom would have an average energy in the amount of $(1/2)kT$ where $k$ is Boltzmann's constant and $T$ is the temperature. Our calculation of the constant-volume heat capacity would be straightforward. Each molecule would be holding, on average, an energy of $(f/2)kT$ where $f$ is the total number of degrees of freedom in the molecule. Note that $Nk = R$ if $N$ is Avogadro's number, which is the case in considering the heat capacity of a mole of molecules. Thus, the total internal energy of the gas would be $(f/2)NkT$ where $N$ is the total number of molecules. The heat capacity (at constant volume) would then be a constant $(f/2)Nk$ the mole-specific heat capacity would be $(f/2)R$ the molecule-specific heat capacity would be $(f/2)k$ and the dimensionless heat capacity would be just $f/2$. Here again, each vibrational degree of freedom contributes 2f. Thus, a mole of nitrous oxide would have a total constant-volume heat capacity (including vibration) of $(13/2)R$ by this calculation.

In summary, the molar heat capacity (mole-specific heat capacity) of an ideal gas with f degrees of freedom is given by

$$C_{V,m} = \frac{f}{2}R$$

This equation applies to all polyatomic gases, if the degrees of freedom are known.

The constant-pressure heat capacity for any gas would exceed this by an extra R. As example $C_p$ would be a total of (15/2)R/mole for nitrous oxide.

## The Effect of Quantum Energy Levels in Storing Energy in Degrees of Freedom

The various degrees of freedom cannot generally be considered to obey classical mechanics, however. Classically, the energy residing in each degree of freedom is assumed to be continuous—it can take on any positive value, depending on the temperature. In reality, the amount of energy that may reside in a particular degree of freedom is quantized: It may only be increased and decreased in finite amounts. A good estimate of the size of this minimum amount is the energy of the first excited state of that degree of freedom above its ground state. For example, the first vibrational state of the hydrogen chloride (HCl) molecule has an energy of about $5.74 \times 10^{-20}$ joule. If this amount of energy were deposited in a classical degree of freedom, it would correspond to a temperature of about 4156 K.

If the temperature of the substance is so low that the equipartition energy of $(1/2)kT$ is much smaller than this excitation energy, then there will be little or no energy in this degree of freedom. This degree of freedom is then said to be "frozen out». As mentioned above, the temperature corresponding to the first excited vibrational state of HCl is about 4156 K. For temperatures well below this value, the vibrational degrees of freedom of the HCl molecule will be frozen out. They will contain little energy and will not contribute to the thermal energy or the heat capacity of HCl gas.

## Energy Storage Mode "Freeze-out" Temperatures

It can be seen that for each degree of freedom there is a critical temperature at which the degree of freedom "unfreezes" and begins to accept energy in a classical way. In the case of translational degrees of freedom, this temperature is that temperature at which the thermal wavelength of the molecules is roughly equal to the size of the container. For a container of macroscopic size (e.g. 10 cm) this temperature is extremely small and has no significance, since the gas will certainly liquify or freeze before this low temperature is reached. For any real gas translational degrees of freedom may be considered to always be classical and contain an average energy of $(3/2)kT$ per molecule.

The rotational degrees of freedom are the next to "unfreeze". In a diatomic gas, for example, the critical temperature for this transition is usually a few tens of kelvins, although with a very light molecule such as hydrogen the rotational energy levels will be spaced so widely that rotational heat capacity may not completely "unfreeze" until considerably higher temperatures are reached. Finally, the vibrational degrees of freedom are generally the last to unfreeze. As an example, for diatomic gases, the critical temperature for the vibrational motion is usually a few thousands of kelvins, and thus for the nitrogen in our example at room temperature, no vibration modes would be excited, and the constant-volume heat capacity at room temperature is $(5/2)R$/mole, not $(7/2)R$/mole. As seen above, with some unusually heavy gases such as iodine gas $I_2$, or bromine gas $Br_2$, some vibrational heat capacity may be observed even at room temperatures.

It should be noted that it has been assumed that atoms have no rotational or internal degrees of freedom. This is in fact untrue. For example, atomic electrons can exist in excited states and even the atomic nucleus can have excited states as well. Each of these internal degrees of freedom are assumed to be frozen out due to their relatively high excitation energy. Nevertheless, for sufficiently high temperatures, these degrees of freedom cannot be ignored. In a few exceptional cases, such molecular electronic transitions are of sufficiently low energy that they contribute to heat capacity at room temperature, or even at cryogenic temperatures. One example of an electronic transition degree of freedom which contributes heat capacity at standard temperature is that of nitric oxide (NO), in which the single electron in an anti-bonding molecular orbital has energy transitions which contribute to the heat capacity of the gas even at room temperature.

An example of a nuclear magnetic transition degree of freedom which is of importance to heat capacity, is the transition which converts the spin isomers of hydrogen gas ($H_2$) into each other. At room temperature, the proton spins of hydrogen gas are aligned 75% of the time, resulting in *orthohydrogen* when they are. Thus, some thermal energy has been stored in the degree of freedom available when *parahydrogen* (in which spins are anti-aligned) absorbs energy, and is converted to the higher energy ortho form. However, at the temperature of liquid hydrogen, not enough heat energy is available to produce orthohydrogen (that is, the transition energy between forms is large enough to "freeze out" at this low temperature), and thus the parahydrogen form predominates. The heat capacity of the transition is sufficient to release enough heat, as orthohydrogen converts to the lower-energy parahydrogen, to boil the hydrogen liquid to gas again, if this evolved heat is not removed with a catalyst after the gas has been cooled and condensed. This example also illustrates the fact that some modes of storage of heat may not be in constant equilibrium with each other in substances, and heat absorbed or released from such phase changes may "catch up" with temperature changes of substances, only after a certain time. In other words, the heat evolved and absorbed from the ortho-para isomeric transition contributes to the heat capacity of hydrogen on long time-scales, but not on *short* time-scales. These time scales may also depend on the presence of a catalyst.

Less exotic phase-changes may contribute to the heat-capacity of substances and systems, as well, as (for example) when water is converted back and forth from solid to liquid or gas form. Phase changes store heat energy entirely in breaking the bonds of the potential energy interactions between molecules of a substance. As in the case of hydrogen, it is also possible for phase changes to be hindered as the temperature drops, thus they do not catch up and become apparent, without a catalyst. For example, it is possible to supercool liquid water to below the freezing point, and not observe the heat evolved when the water changes to ice, so long as the water remains liquid. This heat appears instantly when the water freezes.

## Solid Phase

The dimensionless heat capacity divided by three, as a function of temperature as predicted by the Debye model and by Einstein's earlier model. The horizontal axis is the temperature divided by the Debye temperature. Note that, as expected, the dimensionless heat capacity is zero at absolute zero, and rises to a value of three as the temperature becomes much larger than the Debye temperature. The red line corresponds to the classical limit of the Dulong Petit law

For matter in a crystalline solid phase, the Dulong–Petit law, which was discovered empirically, states that the molar heat capacity assumes the value 3 $R$. Indeed, for solid metallic chemical elements at room temperature, molar heat capacities range from about 2.8 $R$ to 3.4 $R$. Large exceptions at the lower end involve solids composed of relatively low-mass, tightly bonded atoms, such as beryllium at 2.0 $R$, and diamond at only 0.735 $R$. The latter conditions create larger quantum vibrational energy spacing, thus many vibrational modes have energies too high to be populated (and thus are "frozen out") at room temperature. At the higher end of possible heat capacities, heat capacity may exceed $R$ by modest amounts, due to contributions from anharmonic vibrations in solids, and sometimes a modest contribution from conduction electrons in metals. These are not degrees of freedom treated in the Einstein or Debye theories.

The theoretical maximum heat capacity for multi-atomic gases at higher temperatures, as the molecules become larger, also approaches the Dulong–Petit limit of 3 $R$, so long as this is calculated per mole of atoms, not molecules. The reason for this behavior is that, in theory, gases with very large molecules have almost the same high-temperature heat capacity as solids, lacking only the (small) heat capacity contribution that comes from potential energy that cannot be stored between separate molecules in a gas.

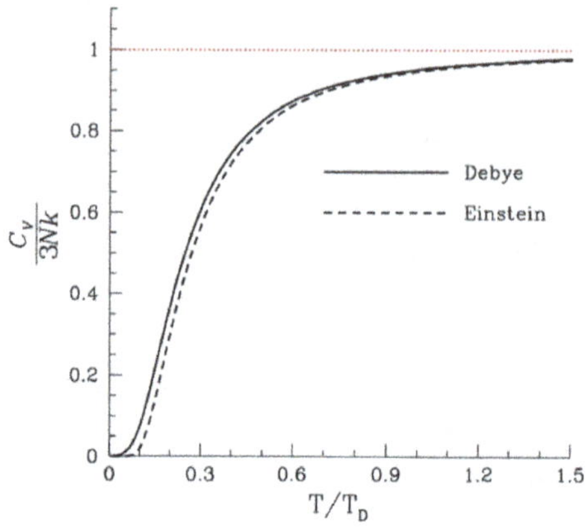


The dimensionless heat capacity divided by three, as a function of temperature as predicted by the Debye model and by Einstein's earlier model. The horizontal axis is the temperature divided by the Debye temperature. Note that, as expected, the dimensionless heat capacity is zero at absolute zero, and rises to a value of three as the temperature becomes much larger than the Debye temperature. The red line corresponds to the classical limit of the Dulong Petit law

The Dulong–Petit limit results from the equipartition theorem, and as such is only valid in the classical limit of a microstate continuum, which is a high temperature limit. For light and non-metallic elements, as well as most of the common molecular solids based on carbon compounds at standard ambient temperature, quantum effects may also play an important role, as they do in multi-atomic gases. These effects usually combine to give heat capacities lower than 3 $R$ per mole of *atoms* in the solid, although in molecular solids, heat capacities calculated *per mole of molecules* in molecular solids may be more than 3 $R$. For example, the heat capacity of water ice at the melting point is about 4.6 $R$ per mole of molecules, but only 1.5 $R$ per mole of atoms. As noted, heat capacity values far lower than 3 $R$ "per atom" (as is the case with diamond and beryllium) result from "freezing out" of possible vibration modes for light atoms at suitably low temperatures, just as happens in many low-mass-atom gases at room temperatures (where vibrational modes are all frozen out). Because of high crystal binding energies, the effects of vibrational mode freezing are observed in solids more often than liquids: for example the heat capacity of liquid water is twice that of ice at near the same temperature, and is again close to the 3 $R$ per mole of atoms of the Dulong–Petit theoretical maximum.

## Liquid Phase

A general theory of the heat capacity of liquids has not yet been achieved, and is still an active area of research. It was long thought that phonon theory is not able to explain the heat capacity of liquids, since liquids only sustain longitudinal, but not transverse phonons, which in solids are responsible for 2/3 of the heat capacity. However, Brillouin scattering experiments with neutrons and with X-rays, confirming an intuition of Yakov Frenkel, have shown that transverse phonons do exist in liquids, albeit restricted to frequencies above a threshold called the Frenkel frequency. Since most energy is contained in these high-frequency modes, a simple modification of the Debye model is sufficient to yield a good approximation to experimental heat capacities of simple liquids.

Amorphous materials can be considered a type of liquid. The specific heat of amorphous materials has characteristic discontinuities at the glass transition temperature. These discontinuities are frequently used to detect the glass transition temperature where a supercooled liquid transforms to a glass.

## Table of Specific Heat Capacities

Note that the especially high molar values, as for paraffin, gasoline, water and ammonia, result from calculating specific heats in terms of moles of *molecules*. If specific heat is expressed per mole of *atoms* for these substances, none of the constant-volume values exceed, to any large extent, the theoretical Dulong–Petit limit of 25 $J{\cdot}mol^{-1}{\cdot}K^{-1}$ = 3 R per mole of atoms. Paraffin, for example, has very large molecules and thus a high heat capacity per mole, but as a substance it does not have remarkable heat capacity in terms of volume, mass, or atom-mol (which is just 1.41 R per mole of atoms, or less than half of most solids, in terms of heat capacity per atom).

In the last column, major departures of solids at standard temperatures from the Dulong–Petit law value of 3 R, are usually due to low atomic weight plus high bond strength (as in diamond) causing some vibration modes to have too much energy to be available to store thermal energy at the measured temperature. For gases, departure from 3 R per mole of atoms in this table is generally due to two factors: (1) failure of the higher quantum-energy-spaced vibration modes in gas molecules to be excited at room temperature, and (2) loss of potential energy degree of freedom for small gas molecules, simply because most of their atoms are not bonded maximally in space to other atoms, as happens in many solids.

| **Notable minima and maxima are shown in bold** | | | | | | |
|---|---|---|---|---|---|---|
| **Table of specific heat capacities at 25°C (298 K) unless otherwise noted. Substance** | **Phase** | **Isobaric mass heat capacity $c_P$ $J{\cdot}g^{-1}{\cdot}K^{-1}$** | **Isobaric molar heat capacity $C_{P,m}$ $J{\cdot}mol^{-1}{\cdot}K^{-1}$** | **Isochore molar heat capacity $C_{V,m}$ $J{\cdot}mol^{-1}{\cdot}K^{-1}$** | **Isobaric volumetric heat capacity $C_{P,v}$ $J{\cdot}cm^{-3}{\cdot}K^{-1}$** | **Isochore atom-molar heat capacity in units of R $C_{V,am}$ $atom{\cdot}mol^{-1}$** |
| Air (Sea level, dry, 0 °C (273.15 K)) | gas | 1.0035 | 29.07 | 20.7643 | 0.001297 | ~ 1.25 R |
| Air (typical room conditions[A]) | gas | 1.012 | 29.19 | 20.85 | 0.00121 | ~ 1.25 R |
| Aluminium | solid | 0.897 | 24.2 | | 2.422 | 2.91 R |
| Ammonia | liquid | 4.700 | 80.08 | | 3.263 | 3.21 R |
| Animal tissue (incl. human) | mixed | 3.5 | | | 3.7* | |
| Antimony | solid | 0.207 | 25.2 | | 1.386 | 3.03 R |
| Argon | gas | 0.5203 | 20.7862 | 12.4717 | | 1.50 R |
| Arsenic | solid | 0.328 | 24.6 | | 1.878 | 2.96 R |
| Beryllium | solid | 1.82 | 16.4 | | 3.367 | 1.97 R |
| Bismuth | solid | 0.123 | 25.7 | | 1.20 | 3.09 R |

| Notable minima and maxima are shown in bold | | | | | | |
|---|---|---|---|---|---|---|
| **Table of specific heat capacities at 25°C (298 K) unless otherwise noted.** <br> **Substance** | **Phase** | **Isobaric mass heat capacity** <br> $c_P$ <br> **J·g⁻¹·K⁻¹** | **Isobaric molar heat capacity $C_{P,m}$** <br> **J·mol⁻¹·K⁻¹** | **Isochore molar heat capacity** <br> $C_{V,m}$ <br> **J·mol⁻¹·K⁻¹** | **Isobaric volumetric heat capacity** <br> $C_{P,v}$ <br> **J·cm⁻³·K⁻¹** | **Isochore atom-molar heat capacity in units of R** <br> $C_{V,am}$ <br> **atom-mol⁻¹** |
| Cadmium | solid | 0.231 | 26.02 | | | 3.13 R |
| Carbon dioxide $CO_2$ | gas | 0.839* | 36.94 | 28.46 | | 1.14 R |
| Chromium | solid | 0.449 | 23.35 | | | 2.81 R |
| Copper | solid | 0.385 | 24.47 | | 3.45 | 2.94 R |
| Diamond | solid | 0.5091 | 6.115 | | 1.782 | 0.74 R |
| Ethanol | liquid | 2.44 | 112 | | 1.925 | 1.50 R |
| Gasoline (octane) | liquid | 2.22 | 228 | | 1.64 | 1.05 R |
| Glass | solid | 0.84 | | | 2.1 | |
| Gold | solid | 0.129 | 25.42 | | 2.492 | 3.05 R |
| Granite | solid | 0.790 | | | 2.17 | |
| Graphite | solid | 0.710 | 8.53 | | 1.534 | 1.03 R |
| Helium | gas | 5.1932 | 20.7862 | 12.4717 | | 1.50 R |
| Hydrogen | gas | 14.30 | 28.82 | | | 1.23 R |
| Hydrogen sulfide $H_2S$ | gas | 1.015* | 34.60 | | | 1.05 R |
| Iron | solid | 0.412 | 25.09 | | 3.537 | 3.02 R |
| Lead | solid | 0.129 | 26.4 | | 1.44 | 3.18 R |
| Lithium | solid | 3.58 | 24.8 | | 1.912 | 2.98 R |
| Lithium at 181 °C | liquid | 4.379 | 30.33 | | 2.242 | 3.65 R |
| Magnesium | solid | 1.02 | 24.9 | | 1.773 | 2.99 R |
| Mercury | liquid | 0.1395 | 27.98 | | 1.888 | 3.36 R |
| Methane at 2 °C | gas | 2.191 | 35.69 | | | 0.85 R |
| Methanol | liquid | 2.14 | 68.62 | | | 1.38 R |
| Molten salt (142–540 °C) | liquid | 1.56 | | | 2.62 | |
| Nitrogen | gas | 1.040 | 29.12 | 20.8 | | 1.25 R |
| Neon | gas | 1.0301 | 20.7862 | 12.4717 | | 1.50 R |
| Oxygen | gas | 0.918 | 29.38 | 21.0 | | 1.26 R |
| Paraffin wax $C_{25}H_{52}$ | solid | 2.5 (ave) | 900 | | 2.325 | 1.41 R |
| Polyethylene (rotomolding grade) | solid | 2.3027 | | | | |
| Silica (fused) | solid | 0.703 | 42.2 | | 1.547 | 1.69 R |
| Silver | solid | 0.233 | 24.9 | | 2.44 | 2.99 R |
| Sodium | solid | 1.230 | 28.23 | | | 3.39 R |
| Steel | solid | 0.466 | | | 3.756 | |

| Notable minima and maxima are shown in bold | | | | | | |
|---|---|---|---|---|---|---|
| **Table of specific heat capacities at 25°C (298 K) unless otherwise noted.**<br>**Substance** | **Phase** | **Isobaric mass heat capacity $c_P$ J·g⁻¹·K⁻¹** | **Isobaric molar heat capacity $C_{P,m}$ J·mol⁻¹·K⁻¹** | **Isochore molar heat capacity $C_{V,m}$ J·mol⁻¹·K⁻¹** | **Isobaric volumetric heat capacity $C_{P,v}$ J·cm⁻³·K⁻¹** | **Isochore atom-molar heat capacity in units of R $C_{V,am}$ atom-mol⁻¹** |
| Tin | solid | 0.227 | 27.112 | | 1.659 | 3.26 R |
| Titanium | solid | 0.523 | 26.060 | | 2.6384 | 3.13 R |
| Tungsten | solid | 0.134 | 24.8 | | 2.58 | 2.98 R |
| Uranium | solid | 0.116 | 27.7 | | 2.216 | 3.33 R |
| Water at 100 °C (steam) | gas | 2.080 | 37.47 | 28.03 | | 1.12 R |
| Water at 25 °C | liquid | 4.1813 | 75.327 | 74.53 | 4.1796 | 3.02 R |
| Water at 100 °C | liquid | 4.1813 | 75.327 | 74.53 | 4.2160 | 3.02 R |
| Water at −10 °C (ice) | solid | 2.05 | 38.09 | | 1.938 | 1.53 R |
| Zinc | solid | 0.387 | 25.2 | | 2.76 | 3.03 R |

[A] Assuming an altitude of 194 metres above mean sea level (the world–wide median altitude of human habitation), an indoor temperature of 23 °C, a dewpoint of 9 °C (40.85% relative humidity), and 760 mm–Hg sea level–corrected barometric pressure (molar water vapor content = 1.16%).

*Derived data by calculation. This is for water-rich tissues such as brain. The whole-body average figure for mammals is approximately 2.9 J·cm⁻³·K⁻¹

## Mass Heat Capacity of Building Materials

(Usually of interest to builders and solar designers)

| Mass heat capacity of building materials | | |
|---|---|---|
| **Substance** | **Phase** | **$c_P$ J·g⁻¹·K⁻¹** |
| Asphalt | solid | 0.920 |
| Brick | solid | 0.840 |
| Concrete | solid | 0.880 |
| Glass, silica | solid | 0.840 |
| Glass, crown | solid | 0.670 |
| Glass, flint | solid | 0.503 |
| Glass, pyrex | solid | 0.753 |
| Granite | solid | 0.790 |
| Gypsum | solid | 1.090 |
| Marble, mica | solid | 0.880 |
| Sand | solid | 0.835 |
| Soil | solid | 0.800 |
| Water | liquid | 4.1813 |
| Wood | solid | 1.7 (1.2 to 2.9) |

# Magnetic Susceptibility

In electromagnetism, the magnetic susceptibility (Latin: *susceptibilis*, "receptive"; denoted χ) is one measure of the magnetic properties of a material. The susceptibility indicates whether a material is attracted into or repelled out of a magnetic field, which in turn has implications for practical applications. Quantitative measures of the magnetic susceptibility also provide insights into the structure of materials, providing insight into bonding and energy levels.

Mathematically it is the ratio of magnetization *I* (magnetic moment per unit volume) to the applied magnetizing field intensity *H*.

## Definition of Volume Susceptibility

Magnetic susceptibility is a dimensionless proportionality constant that indicates the degree of magnetization of a material in response to an applied magnetic field. A related term is magnetizability, the proportion between magnetic moment and magnetic flux density. A closely related parameter is the permeability, which expresses the total magnetization of material and volume.

The *volume magnetic susceptibility*, represented by the symbol $\chi_v$ (often simply $\chi$, sometimes $\chi_m$ – magnetic, to distinguish from the electric susceptibility), is defined in the International System of Units — in other systems there may be additional constants — by the following relationship:

$$\mathrm{M} = \chi_v \mathrm{H}.$$

Here

M is the magnetization of the material (the magnetic dipole moment per unit volume), measured in amperes per meter, and

H is the magnetic field strength, also measured in amperes per meter.

$\chi_v$ is therefore a dimensionless quantity.

Using SI units, the magnetic induction B is related to H by the relationship

$$\mathrm{B} = \mu_0(\mathrm{H}+\mathrm{M}) = \mu_0(1+\chi_v)\mathrm{H} = \mu\mathrm{H}$$

where $\mu_0$ is the magnetic constant, and $(1+\chi_v)$ is the relative permeability of the material. Thus the *volume magnetic susceptibility* $\chi_v$ and the magnetic permeability $\mu$ are related by the following formula:

$$\mu = \mu_0(1+\chi_v).$$

Sometimes an auxiliary quantity called *intensity of magnetization* (also referred to as *magnetic polarisation* J) and measured in teslas, is defined as

$$\mathrm{I} = \mu_0 \mathrm{M}.$$

This allows an alternative description of all magnetization phenomena in terms of the quantities I and B, as opposed to the commonly used M and H.

Note that these definitions are according to SI conventions. However, many tables of magnetic susceptibility give CGS values (more specifically emu-cgs, short for electromagnetic units, or Gaussian-cgs; both are the same in this context). These units rely on a different definition of the permeability of free space:

$$B^{cgs} = H^{cgs} + 4\pi M^{cgs} = \left(1 + 4\pi\chi_v^{cgs}\right)H^{cgs}$$

The dimensionless CGS value of volume susceptibility is multiplied by 4π to give the dimensionless SI volume susceptibility value:

$$\chi_v^{SI} = 4\pi\chi_v^{cgs}$$

For example, the CGS volume magnetic susceptibility of water at 20 °C is $-7.19\times10^{-7}$ which is $-9.04\times10^{-6}$ using the SI convention.

In physics it is common (in older literature) to see CGS mass susceptibility given in emu/g, so to convert to SI volume susceptibility we use the conversion

$$\chi_v^{SI} = 4\pi\rho^{cgs}\chi_m^{cgs}$$

where $\rho^{cgs}$ is the density given in g/cm³, or

$$\chi_v^{SI} = \left(4\pi\times10^{-3}\right)\rho^{SI}\chi_m^{cgs}$$

where $\rho^{SI}$ is the density given in kg/m³.

## Mass Susceptibility and Molar Susceptibility

There are two other measures of susceptibility, the *mass magnetic susceptibility* ($\chi_{mass}$ or $\chi_g$, sometimes $\chi_m$), measured in $m^3\cdot kg^{-1}$ in SI or in $cm^3\cdot g^{-1}$ in CGS and the *molar magnetic susceptibility* ($\chi_{mol}$) measured in $m^3\cdot mol^{-1}$ (SI) or $cm^3\cdot mol^{-1}$ (CGS) that are defined below, where ρ is the density in $kg\cdot m^{-3}$ (SI) or $g\cdot cm^{-3}$ (CGS) and M is molar mass in $kg\cdot mol^{-1}$ (SI) or $g\cdot mol^{-1}$ (CGS).

$$\chi_{mass} = \frac{\chi_v}{\rho}$$

$$\chi_{mol} = M\chi_{mass} = \frac{M\chi_v}{\rho}$$

## Sign of Susceptibility: Diamagnetics and other Types of Magnetism

If χ is positive, a material can be paramagnetic. In this case, the magnetic field in the material is strengthened by the induced magnetization. Alternatively, if χ is negative, the material is diamagnetic. In this case, the magnetic field in the material is weakened by the induced magnetization. Generally, non-magnetic materials are said to be para- or diamagnetic because they do not possess permanent magnetization without external magnetic field. Ferromagnetic, ferrimagnetic, or anti-ferromagnetic materials have positive susceptibility and possess permanent magnetization even without external magnetic field.

## Experimental Methods to Determine Susceptibility

Volume magnetic susceptibility is measured by the force change felt upon a substance when a magnetic field gradient is applied. Early measurements are made using the Gouy balance where a sample is hung between the poles of an electromagnet. The change in weight when the electromagnet is turned on is proportional to the susceptibility. Today, high-end measurement systems use a superconductive magnet. An alternative is to measure the force change on a strong compact magnet upon insertion of the sample. This system, widely used today, is called the Evans balance. For liquid samples, the susceptibility can be measured from the dependence of the NMR frequency of the sample on its shape or orientation. Another method using MRI/NMR techniques measures the magnetic field distortion around a sample immersed in water inside an MR scanner. This method is highly accurate for diamagnetic materials with susceptibilities similar to water.

## Tensor Susceptibility

The magnetic susceptibility of most crystals is not a scalar quantity. Magnetic response M is dependent upon the orientation of the sample and can occur in directions other than that of the applied field H. In these cases, volume susceptibility is defined as a tensor

$$M_i = H_j \chi_{ij}$$

where *i* and *j* refer to the directions (e.g., *x* and *y* in Cartesian coordinates) of the applied field and magnetization, respectively. The tensor is thus rank 2 (second order), dimension (3,3) describing the component of magnetization in the *i*-th direction from the external field applied in the *j*-th direction.

## Differential Susceptibility

In ferromagnetic crystals, the relationship between M and H is not linear. To accommodate this, a more general definition of *differential susceptibility* is used

$$\chi_{ij}^d = \frac{\partial M_i}{\partial H_j}$$

where $\chi_{ij}^d$ is a tensor derived from partial derivatives of components of M with respect to components of H. When the coercivity of the material parallel to an applied field is the smaller of the two, the differential susceptibility is a function of the applied field and self interactions, such as the magnetic anisotropy. When the material is not saturated, the effect will be nonlinear and dependent upon the domain wall configuration of the material.

## Susceptibility in the Frequency Domain

When the magnetic susceptibility is measured in response to an AC magnetic field (i.e. a magnetic field that varies sinusoidally), this is called *AC susceptibility*. AC susceptibility (and the closely related "AC permeability") are complex number quantities, and various phenomena (such as resonances) can be seen in AC susceptibility that cannot in constant-field (DC) susceptibility. In particular, when an AC field is applied perpendicular to the detection direction (called the "transverse susceptibility" regardless of the frequency), the effect has a peak at the ferromagnetic resonance frequency of the material with a given static applied field. Currently, this effect is called the *microwave permeability* or *network ferromagnetic resonance* in the literature. These results are sensitive to the domain wall configuration of the material and eddy currents.

In terms of ferromagnetic resonance, the effect of an ac-field applied along the direction of the magnetization is called *parallel pumping*.

### Sources of Confusion in Published Data

The CRC Handbook of Chemistry and Physics has one of the only published magnetic susceptibility tables. Some of the data (e.g., for Al, Bi, and diamond) is listed as CGS. CGS has caused confusion to some readers. CGS is an abbreviation of *centimeters–grams–seconds*; it represents the form of the units, but CGS does not specify units. Correct units of magnetic susceptibility in CGS is $cm^3/mol$ or $cm^3/g$. Molar susceptibility and mass susceptibility are both listed in the CRC. Some table have listed magnetic susceptibility of diamagnets as positives. It is important to check the header of the table for the correct units and sign of magnetic susceptibility readings.

### Magnetic susceptibility

The magnetic (spin) susceptibility of a non-interacting electron gas is called the Pauli spin susceptibility and is given by $\chi = \mu_B^2 g(\epsilon_F)$ where $\mu_B$ is the Bohr magneton. Since the temperatures at which the magnetic susceptibility is typically measured are much smaller than $\frac{\epsilon_F}{k_B}$, it is independent of temperature. The Pauli susceptibility has a typical value of $10^{-6}\,\frac{\text{cm}^3}{\text{mole}}$. In addition, there are other contributions to the susceptibility namely, the Van Vleck orbital paramagnetism, the diamagnetism from the orbital motion of the core electrons, and Landau diamagnetism from the orbital motion of the free electrons. The magnetic susceptibility of conventional metals is independent of temperature.

## Hall Effect

The Hall effect is the production of a voltage difference (the Hall voltage) across an electrical conductor, transverse to an electric current in the conductor and to an applied magnetic field perpendicular to the current. It was discovered by Edwin Hall in 1879.

The Hall coefficient is defined as the ratio of the induced electric field to the product of the current density and the applied magnetic field. It is a characteristic of the material from which the conductor is made, since its value depends on the type, number, and properties of the charge carriers that constitute the current.

## Discovery

The Hall effect was discovered in 1879 by Edwin Hall while he was working on his doctoral degree at Johns Hopkins University in Baltimore, Maryland. His measurements of the tiny effect produced in the apparatus he used were an experimental tour de force, accomplished 18 years before the electron was discovered and published under the name "On a New Action of the Magnet on Electric Currents".

## Theory

The Hall effect is due to the nature of the current in a conductor. Current consists of the movement of many small charge carriers, typically electrons, holes, ions or all three. When a magnetic field is present, these charges experience a force, called the Lorentz force. When such a magnetic field is absent, the charges follow approximately straight, 'line of sight' paths between collisions with impurities, phonons, etc. However, when a magnetic field with a perpendicular component is applied, their paths between collisions are curved, thus moving charges accumulate on one face of the material. This leaves equal and opposite charges exposed on the other face, where there is a scarcity of mobile charges. The result is an asymmetric distribution of charge density across the Hall element, arising from a force that is perpendicular to both the 'line of sight' path and the applied magnetic field. The separation of charge establishes an electric field that opposes the migration of further charge, so a steady electrical potential is established for as long as the charge is flowing.

In classical electromagnetism electrons move in the opposite direction of the current $I$ (by convention "current" describes a theoretical "hole flow"). In some semiconductors it *appears* "holes" are actually flowing because the direction of the voltage is opposite to the derivation below.

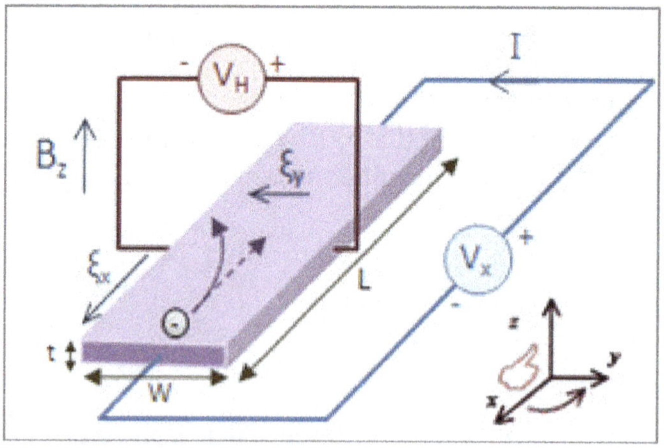


Hall Effect measurement setup for electrons. Initially, the electrons follow the curved arrow, due to the magnetic force. At some distance from the current-introducing contacts, electrons pile up on the left side and deplete from the right side, which creates an electric field $\xi_y$ in the direction of the assigned $V_H$. $V_H$ is negative for some semi-conductors where "holes" appear to flow. In steady-state, $\xi_y$ will be strong enough to exactly cancel out the magnetic force, thus the electrons follow the straight arrow (dashed).

For a simple metal where there is only one type of charge carrier (electrons), the Hall voltage $V_H$ can be derived by using the Lorentz force and seeing that, in the steady-state condition, charges are not moving in the y-axis direction - the magnetic force on each electron in the y-axis direction is cancelled by a y-axis electrical force due to the buildup of charges. The $v_x$ term is the drift velocity of the current which is assumed at this point to be holes by convention. The $v_x B_z$ term is negative in the y-axis direction by the right hand rule.

$$\mathrm{F} = q\left[\mathrm{E} + (\mathrm{v} \times \mathrm{B})\right]$$

$0 = E_y - v_x B_z$ where $E_y$ is assigned in direction of y-axis, not with the arrow as in the image.

In wires, electrons instead of holes are flowing, so $v_x \to -v_x$ and $q \to -q$. Also $E_y = \frac{-V_H}{w}$. Substituting these changes gives

$$V_H = v_x B_z w$$

The conventional "hole" current is in the negative direction of the electron current and the negative of the electrical charge which gives $I_x = ntw(-v_x)(-e)$ where $n$ is charge carrier density, $tw$ is the cross-sectional area, and $-e$ is the charge of each electron. Solving for $w$ and plugging into the above gives the Hall voltage:

$$V_H = \frac{I_x B_z}{nte}$$

If the charge build up had been positive (as it appears in some semiconductors), then the $V_H$ assigned in the image would have been negative (positive charge would have built up on the left side).

The Hall coefficient is defined as

$$R_H = \frac{E_y}{j_x B_z}$$

where $j$ is the current density of the carrier electrons, and $E_y$ is the induced electric field. In SI units, this becomes

$$R_H = \frac{E_y}{j_x B} = \frac{V_H t}{IB} = -\frac{1}{ne}.$$

(The units of $R_{\mathrm{H}}$ are usually expressed as m³/C, or Ω·cm/G, or other variants.) As a result, the Hall effect is very useful as a means to measure either the carrier density or the magnetic field.

One very important feature of the Hall effect is that it differentiates between positive charges moving in one direction and negative charges moving in the opposite. The Hall effect offered the first real proof that electric currents in metals are carried by moving electrons, not by protons. The Hall effect also showed that in some substances (especially p-type semiconductors), it is more appropriate to think of the current as positive "holes" moving rather than negative electrons. A common source of confusion with the Hall Effect is that holes moving to the left are really electrons moving to the right, so one expects the same sign of the Hall coefficient for both electrons and holes. This confusion, however, can only be resolved by modern quantum mechanical theory of transport in solids.

The sample inhomogeneity might result in spurious sign of the Hall effect, even in ideal van der Pauw configuration of electrodes. For example, positive Hall effect was observed in evidently n-type semiconductors. Another source of artifact, in uniform materials, occurs when the sample's aspect ratio is not long enough: the full Hall voltage only develops far away from the current-introducing contacts, since at the contacts the transverse voltage is shorted out to zero.

## Hall Effect in Semiconductors

When a current-carrying semiconductor is kept in a magnetic field, the charge carriers of the semiconductor experience a force in a direction perpendicular to both the magnetic field and the current. At equilibrium, a voltage appears at the semiconductor edges.

The simple formula for the Hall coefficient given above becomes more complex in semiconductors where the carriers are generally both electrons and holes which may be present in different concentrations and have different mobilities. For moderate magnetic fields the Hall coefficient is

$$R_H = \frac{p\mu_h^2 - n\mu_e^2}{e(p\mu_h + n\mu_e)^2}$$

or equivalently

$$R_H = \frac{(p - nb^2)}{e(p + nb)^2}$$

with

$$b = \frac{\mu_e}{\mu_h}.$$

Here $n$ is the electron concentration, $p$ the hole concentration, $\mu_e$ the electron mobility, $\mu_h$ the hole mobility and $e$ the elementary charge. For large applied fields the simpler expression analogous to that for a single carrier type holds.

## Relationship with Star Formation

Although it is well known that magnetic fields play an important role in star formation, research models indicate that Hall diffusion critically influences the dynamics of gravitational collapse that forms protostars.

## Quantum Hall Effect

For a two-dimensional electron system which can be produced in a MOSFET, in the presence of large magnetic field strength and low temperature, one can observe the quantum Hall effect, in which the Hall conductance σ undergoes quantum Hall transitions to take on the quantized values.

## Spin Hall Effect

The spin Hall effect consists in the spin accumulation on the lateral boundaries of a current-carrying sample. No magnetic field is needed. It was predicted by M. I. Dyakonov and V. I. Perel in 1971 and observed experimentally more than 30 years later, both in semiconductors and in metals, at cryogenic as well as at room temperatures.

## Quantum Spin Hall Effect

For mercury telluride two dimensional quantum wells with strong spin-orbit coupling, in zero magnetic field, at low temperature, the Quantum spin Hall effect has been recently observed.

## Anomalous Hall Effect

In ferromagnetic materials (and paramagnetic materials in a magnetic field), the Hall resistivity includes an additional contribution, known as the anomalous Hall effect (or the extraordinary Hall effect), which depends directly on the magnetization of the material, and is often much larger than the ordinary Hall effect. (Note that this effect is *not* due to the contribution of the magnetization to the total magnetic field.) For example, in nickel, the anomalous Hall coefficient is about 100 times larger than the ordinary Hall coefficient near the Curie temperature, but the two are similar at very low temperatures. Although a well-recognized phenomenon, there is still debate about its origins in the various materials. The anomalous Hall effect can be either an *extrinsic* (disorder-related) effect due to spin-dependent scattering of the charge carriers, or an *intrinsic* effect which can be described in terms of the Berry phase effect in the crystal momentum space (*k*-space).

## Hall Effect in Ionized Gases

The Hall effect in an ionized gas (plasma) is significantly different from the Hall effect in solids (where the Hall parameter is always very inferior to unity). In a plasma, the Hall parameter can take any value. The Hall parameter, $\beta$, in a plasma is the ratio between the electron gyrofrequency, $\Omega_e$, and the electron-heavy particle collision frequency, $\nu$:

$$\beta = \frac{\Omega_e}{\nu} = \frac{eB}{m_e \nu}$$

where

$e$ is the elementary charge (approx. $1.6 \times 10^{-19}$ C)

$B$ is the magnetic field (in teslas)

$m_e$ is the electron mass (approx. $9.1 \times 10^{-31}$ kg).

The Hall parameter value increases with the magnetic field strength.

Physically, the trajectories of electrons are curved by the Lorentz force. Nevertheless, when the Hall parameter is low, their motion between two encounters with heavy particles (neutral or ion) is almost linear. But if the Hall parameter is high, the electron movements are highly curved. The current density vector, $J$, is no longer colinear with the electric field vector, $E$. The two vectors $J$ and $E$ make the Hall angle, $\theta$, which also gives the Hall parameter:

$$\beta = \tan(\theta).$$

## Applications

Hall probes are often used as magnetometers, i.e. to measure magnetic fields, or inspect materials (such as tubing or pipelines) using the principles of magnetic flux leakage.

Hall effect devices produce a very low signal level and thus require amplification. While suitable for laboratory instruments, the vacuum tube amplifiers available in the first half of the 20th century were too expensive, power consuming, and unreliable for everyday applications. It was only with the development of the low cost integrated circuit that the Hall effect sensor became suitable for mass application. Many devices now sold as Hall effect sensors in fact contain both the sensor as described above plus a high gain integrated circuit (IC) amplifier in a single package. Recent advances have further added into one package an analog-to-digital converter and $I^2C$ (Inter-integrated circuit communication protocol) IC for direct connection to a microcontroller's I/O port.

## Advantages Over other Methods

Hall effect devices (when appropriately packaged) are immune to dust, dirt, mud, and water. These characteristics make Hall effect devices better for position sensing than alternative means such as optical and electromechanical sensing.

Hall effect current sensor with internal integrated circuit amplifier. 8 mm opening. Zero current output voltage is midway between the supply voltages that maintain a 4 to 8 Volt differential. Non-zero current response is proportional to the voltage supplied and is linear to 60 amperes for this particular (25 A) device.

When electrons flow through a conductor, a magnetic field is produced. Thus, it is possible to create a non-contacting current sensor. The device has three terminals. A sensor voltage is applied across two terminals and the third provides a voltage proportional to the current being sensed. This has several advantages; no additional resistance (a *shunt*, required for the most common current sensing method) need be inserted in the primary circuit. Also, the voltage present on the line to be sensed is not transmitted to the sensor, which enhances the safety of measuring equipment.

## Disadvantages Compared with other methods

Magnetic flux from the surroundings (such as other wires) may diminish or enhance the field the Hall probe intends to detect, rendering the results inaccurate. Also, as Hall voltage is often on the order of millivolts, the output from this type of sensor cannot be used to directly drive actuators but instead must be amplified by a transistor-based circuit.

Ways to measure component positions within an electromagnetic system, such as a brushless direct current motor, include (I.) the Hall Effect, (II.) light detection with a light-dark position encoder such as a Gray code disk and (III.) induced voltage by moving the amount of metal core inserted into a transformer. When Hall is compared to photo-sensitive methods, it is harder to get absolute position with Hall. Hall detection is also sensitive to stray magnetic fields.

## Contemporary Applications

Hall effect sensors are readily available from a number of different manufacturers, and may be used in various sensors such as rotating speed sensors (bicycle wheels, gear-teeth, automotive speedometers, electronic ignition systems), fluid flow sensors, current sensors, and pressure sensors. Common applications are often found where a robust and contactless switch or potentiometer is required. These include: electric airsoft guns, triggers of electropneumatic paintball guns, go-cart speed controls, smart phones, and some global positioning systems.

## Ferrite Toroid Hall Effect Current Transducer

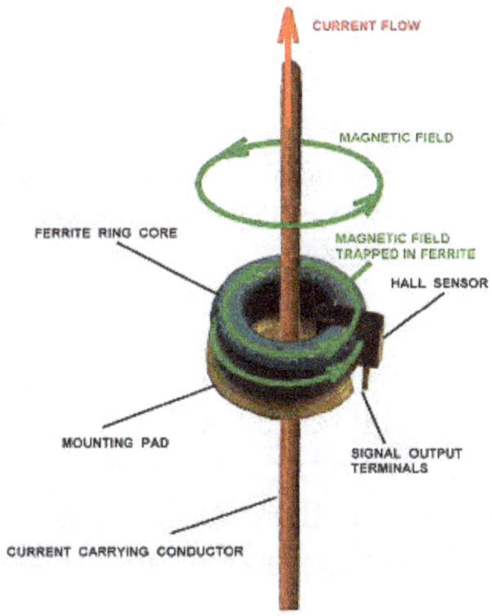


Diagram of Hall effect current transducer integrated into ferrite ring.

Hall sensors can detect stray magnetic fields easily, including that of Earth, so they work well as electronic compasses: but this also means that such stray fields can hinder accurate measurements of small magnetic fields. To solve this problem, Hall sensors are often integrated with magnetic shielding of some kind. For example, a Hall sensor integrated into a ferrite ring (as shown) can reduce the detection of stray fields by a factor of 100 or better (as the external magnetic fields cancel across the ring, giving no residual magnetic flux). This configuration also provides an improvement in signal-to-noise ratio and drift effects of over 20 times that of a bare Hall device. The range of a given feedthrough sensor may be extended upward and downward by appropriate wiring. To extend the range to lower currents, multiple turns of the current-carrying wire may be made through the opening, each turn adding to the sensor output the same quantity; when the sensor is installed onto a printed circuit board, the turns can be carried out by a staple on the board. To extend the range to higher currents, a current divider may be used. The divider splits the current across two wires of differing widths and the thinner wire, carrying a smaller proportion of the total current, passes through the sensor.

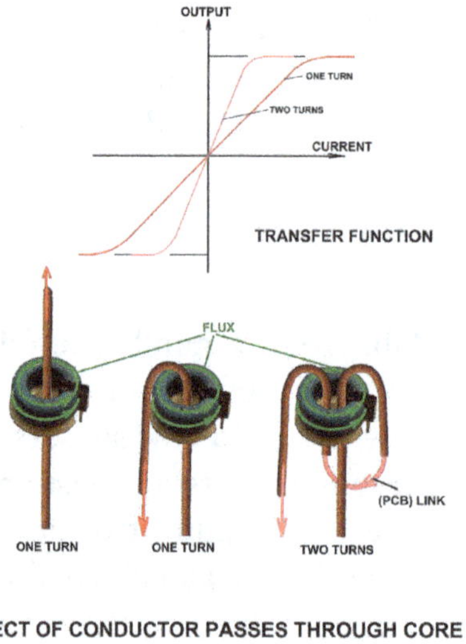


Multiple 'turns' and corresponding transfer function.

## Split Ring Clamp-on Sensor

A variation on the ring sensor uses a split sensor which is clamped onto the line enabling the device to be used in temporary test equipment. If used in a permanent installation, a split sensor allows the electric current to be tested without dismantling the existing circuit.

## Analog Multiplication

The output is proportional to both the applied magnetic field and the applied sensor voltage. If the magnetic field is applied by a solenoid, the sensor output is proportional to the product of the current through the solenoid and the sensor voltage. As most applications requiring computation are now performed by small digital computers, the remaining useful application is in power sensing, which combines current sensing with voltage sensing in a single Hall effect device.

## Power Measurement

By sensing the current provided to a load and using the device's applied voltage as a sensor voltage it is possible to determine the power dissipated by a device.

## Position and Motion Sensing

Hall effect devices used in motion sensing and motion limit switches can offer enhanced reliability in extreme environments. As there are no moving parts involved within the sensor or magnet, typical life expectancy is improved compared to traditional electromechanical switches. Additionally, the sensor and magnet may be encapsulated in an appropriate protective material. This application is used in brushless DC motors.

## Automotive Ignition and Fuel Injection

Commonly used in distributors for ignition timing (and in some types of crank and camshaft position sensors for injection pulse timing, speed sensing, etc.) the Hall effect sensor is used as a direct replacement for the mechanical breaker points used in earlier automotive applications. Its use as an ignition timing device in various distributor types is as follows. A stationary permanent magnet and semiconductor Hall effect chip are mounted next to each other separated by an air gap, forming the Hall effect sensor. A metal rotor consisting of windows and tabs is mounted to a shaft and arranged so that during shaft rotation, the windows and tabs pass through the air gap between the permanent magnet and semiconductor Hall chip. This effectively shields and exposes the Hall chip to the permanent magnet's field respective to whether a tab or window is passing though the Hall sensor. For ignition timing purposes, the metal rotor will have a number of equal-sized tabs and windows matching the number of engine cylinders. This produces a uniform square wave output since the on/off (shielding and exposure) time is equal. This signal is used by the engine computer or ECU to control ignition timing. Many automotive Hall effect sensors have a built-in internal NPN transistor with an open collector and grounded emitter, meaning that rather than a voltage being produced at the Hall sensor signal output wire, the transistor is turned on providing a circuit to ground through the signal output wire.

## Wheel Rotation Sensing

The sensing of wheel rotation is especially useful in anti-lock braking systems. The principles of such systems have been extended and refined to offer more than anti-skid functions, now providing extended vehicle handling enhancements.

## Electric Motor Control

Some types of brushless DC electric motors use Hall effect sensors to detect the position of the rotor and feed that information to the motor controller. This allows for more precise motor control.

## Industrial applications

Applications for Hall Effect sensing have also expanded to industrial applications, which now use Hall Effect joysticks to control hydraulic valves, replacing the traditional mechanical levers with contactless sensing. Such applications include mining trucks, backhoe loaders, cranes, diggers, scissor lifts, etc.

## Spacecraft Propulsion

A Hall-effect thruster (HET) is a relatively low power device that is used to propel some spacecraft, after it gets into orbit or farther out into space. In the HET, atoms are ionized and accelerated by an electric field. A radial magnetic field established by magnets on the thruster is used to trap electrons which then orbit and create an electric field due to the Hall effect. A large potential is established between the end of the thruster where neutral propellant is fed, and the part where electrons are produced; so, electrons trapped in the magnetic field cannot drop to the lower potential. They are thus extremely energetic, which means that they can ionize neutral atoms. Neutral propellant is pumped into the chamber and is ionized by the trapped electrons. Positive ions and electrons are then ejected from the thruster as a quasineutral plasma, creating thrust.

## The Corbino Effect

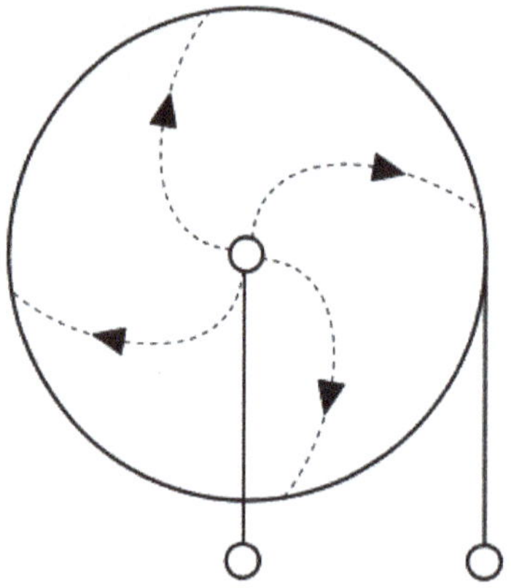

Corbino disc – dashed curves represent logarithmic spiral paths of deflected electrons.

The Corbino effect is a phenomenon involving the Hall effect, but a disc-shaped metal sample is used in place of a rectangular one. Because of its shape the Corbino disc allows the observation of Hall effect–based magnetoresistance without the associated Hall voltage.

A radial current through a circular disc, subjected to a magnetic field perpendicular to the plane of the disc, produces a "circular" current through the disc.

The absence of the free transverse boundaries renders the interpretation of the Corbino effect simpler than that of the Hall effect.

The Hall effect is used to determine the concentration and nature of charge carriers in a material. In the standard Hall effect geometry, a magnetic field is applied perpendicular to the direction in which an ohmic current is flowing. Due to the Lorentz force on the charge carriers, a voltage develops along the third orthogonal direction and is called the Hall voltage. The Hall coefficient is defined as $R_H = \frac{E_y}{j_x B_z}$ where $j_x$ is the ohmic current density in the $x$-direction, $B_z$ is the applied magnetic field in the $z$-direction, and $E_y$ is the electric field that is developed in the $y$-direction. The Hall coefficient is equal to $-\frac{1}{ne}$ (MKS units) if the charge carriers are electrons (of charge e and density n). A typical value of the Hall coefficient in metals is $10^{-10}\,\frac{\mathrm{m}^3}{\mathrm{C}}$. In contrast to the above, there are qualitative changes that take place in the properties of materials when they become superconducting.

## References


- R. M. Pashley; M. Rzechowicz; L. R. Pashley; M. J. Francis (2005). "De-Gassed Water is a Better Cleaning Agent". The Journal of Physical Chemistry B. 109 (3): 1231–8. PMID 16851085. doi:10.1021/jp045975a
- John C. Gallop (1990). SQUIDS, the Josephson Effects and Superconducting Electronics. CRC Press. pp. 3, 20. ISBN 0-7503-0051-5
- Stephenson, C.; Hubler, A. (2015). "Stability and conductivity of self assembled wires in a transverse electric field". Sci.Rep.5. 8323. doi:10.1038/srep15044
- Fraundorf, P. (2003). "Heat capacity in bits". American Journal of Physics. 71 (11): 1142. Bibcode:2003AmJPh..71.1142F. arXiv:cond-mat/9711074. doi:10.1119/1.1593658
- Keith Welch. "Questions & Answers - How do you explain electrical resistance?". Thomas Jefferson National Accelerator Facility. Retrieved 28 April 2017
- Hugh O. Pierson, Handbook of carbon, graphite, diamond, and fullerenes: properties, processing, and applications, p. 61, William Andrew, 1993 ISBN 0-8155-1339-9
- Hogan, C. (1969). "Density of States of an Insulating Ferromagnetic Alloy". Physical Review. 188 (2): 870. Bibcode:1969PhRv..188..870H. doi:10.1103/PhysRev.188.870
- Bennett, L. H.; Page, C. H. & Swartzendruber, L. J. (1978). "Comments on units in magnetism". Journal of Research of the National Bureau of Standards. NIST, USA. 83 (1): 9–12
- International Bureau of Weights and Measures (2006), The International System of Units (SI) (PDF) (8th ed.), ISBN 92-822-2213-6
- Ojovan, Michael I. (2008). "Viscosity and Glass Transition in Amorphous Oxides". Advances in Condensed Matter Physics. 2008: 1. Bibcode:2008AdCMP2008....1O. doi:10.1155/2008/817829
- Physicists Show Electrons Can Travel More Than 100 Times Faster in Graphene Archived September 19, 2013, at the Wayback Machine.. Newsdesk.umd.edu (2008-03-24). Retrieved on 2014-02-03
- N. A. Sinitsyn (2008). "Semiclassical Theories of the Anomalous Hall Effect". Journal of Physics: Condensed Matter. 20 (2): 023201. Bibcode:2008JPCM...20b3201S. arXiv:0712.0183. doi:10.1088/0953-8984/20/02/023201
- Adams, E. P. (1915). "The Hall and Corbino effects". Proceedings of the American Philosophical Society. American Philosophical Society. 54 (216): 47–51. ISBN 978-1-4223-7256-2. Retrieved 2009-01-24
- Robert Karplus and J. M. Luttinger (1954). "Hall Effect in Ferromagnetics". Phys. Rev. 95 (5): 1154–1160. Bibcode:1954PhRv...95.1154K. doi:10.1103/PhysRev.95.1154
- Edwin Hall (1879). "On a New Action of the Magnet on Electric Currents". American Journal of Mathematics. 2 (3): 287–92. JSTOR 2369245. doi:10.2307/2369245. Archived from the original on 2011-07-27. Retrieved 2008-02-28

# Permissions

All chapters in this book are published with permission under the Creative Commons Attribution Share Alike License or equivalent. Every chapter published in this book has been scrutinized by our experts. Their significance has been extensively debated. The topics covered herein carry significant information for a comprehensive understanding. They may even be implemented as practical applications or may be referred to as a beginning point for further studies.

We would like to thank the editorial team for lending their expertise to make the book truly unique. They have played a crucial role in the development of this book. Without their invaluable contributions this book wouldn't have been possible. They have made vital efforts to compile up to date information on the varied aspects of this subject to make this book a valuable addition to the collection of many professionals and students.

This book was conceptualized with the vision of imparting up-to-date and integrated information in this field. To ensure the same, a matchless editorial board was set up. Every individual on the board went through rigorous rounds of assessment to prove their worth. After which they invested a large part of their time researching and compiling the most relevant data for our readers.

The editorial board has been involved in producing this book since its inception. They have spent rigorous hours researching and exploring the diverse topics which have resulted in the successful publishing of this book. They have passed on their knowledge of decades through this book. To expedite this challenging task, the publisher supported the team at every step. A small team of assistant editors was also appointed to further simplify the editing procedure and attain best results for the readers.

Apart from the editorial board, the designing team has also invested a significant amount of their time in understanding the subject and creating the most relevant covers. They scrutinized every image to scout for the most suitable representation of the subject and create an appropriate cover for the book.

The publishing team has been an ardent support to the editorial, designing and production team. Their endless efforts to recruit the best for this project, has resulted in the accomplishment of this book. They are a veteran in the field of academics and their pool of knowledge is as vast as their experience in printing. Their expertise and guidance has proved useful at every step. Their uncompromising quality standards have made this book an exceptional effort. Their encouragement from time to time has been an inspiration for everyone.

The publisher and the editorial board hope that this book will prove to be a valuable piece of knowledge for students, practitioners and scholars across the globe.

# Index

www.ingramcontent.com/pod-product-compliance
Lightning Source LLC
Chambersburg PA
CBHW082042190326
41458CB00010B/3437